Foundation Engineering

Foundation Engineering

Ivy Baxter

NY RESEARCH
P R E S S

New York

Published by NY Research Press
118-35 Queens Blvd., Suite 400,
Forest Hills, NY 11375, USA
www.nyresearchpress.com

Foundation Engineering
Ivy Baxter

International Standard Book Number: 978-1-64725-422-3 (Hardback)

Cataloging-in-Publication Data

Foundation engineering / Ivy Baxter.
   p. cm.
Includes bibliographical references and index.
ISBN 978-1-64725-422-3
1. Foundations. 2. Structural engineering. I. Baxter, Ivy.
TA775 .P75 2023
624.15--dc23

Contents

Permissions

Index

Preface

A foundation refers to the part of the structure that bears the load of the superstructure (structure above the plinth level) and also transfers the load from the structure to the ground/ soil. Foundation engineering is a branch of civil engineering involved in the study and design of structures below the plinth level. It deals with the application of soil mechanics and rock mechanics in the design of foundation elements of structures. The design and construction of a foundation must have the power to sustain and transmit the dead and imposed loads to the soil. Foundations can be classified into four types including shallow foundations, deep foundations, monopile foundations, and individual footings. A foundation that is constructed where the hard strata are extremely near the ground surface rather than deep down the sub-layers of the ground, or the quality of the topsoil is sufficient for bearing is termed as shallow foundation. Continuous wall footings and pad or mat footings are common types of shallow foundations. This book provides comprehensive insights into the field of foundation engineering. It will serve a valuable source of reference for students interested in this branch of civil engineering.

Various studies have approached the subject by analyzing it with a single perspective, but the present book provides diverse methodologies and techniques to address this field. This book contains theories and applications needed for understanding the subject from different perspectives. The aim is to keep the readers informed about the progresses in the field; therefore, the contributions were carefully examined to compile novel researches by specialists from across the globe.

Indeed, the job of the editor is the most crucial and challenging in compiling all chapters into a single book. In the end, I would extend my sincere thanks to the chapter authors for their profound work. I am also thankful for the support provided by my family and colleagues during the compilation of this book.

Ivy Baxter

Site Investigation and Selection of Foundation

1.1　Scope and Objectives

Soil is used as:

- Construction material.

- Supporting material for carrying the loads of the super-structure through their foundations.

The function of a properly designed foundation is to support loads resting on it without causing excessive stresses within the soil mass at any depth of earth foundation.

Stresses are considered excessive if the complete rupture within the soil mass occurs or if detrimental settlements result. It is apparent that one of the very important steps and solution of a foundation problem is determining underground conditions that will affect the design.

The field and laboratory investigations required to obtain necessary information about geology, hydrology and soil conditions. A geotechnical property of soil at the prospective building site and the performance of different soil types encountered when acted upon by structural loads, water and temperature are known as sub-surface investigations or soil exploration programme.

The principals and objectives for a Highway design site investigation are as follows:

- Design: Obtain all the design parameters necessary for the works.

- Suitability: The site and surroundings suitable for the highway.

- Construction: There are any potential ground or ground water conditions that would affect the construction?

- Materials: There are any materials available on the site, what quantity and quality?

- Effect of changes: How will the design affect adjacent properties and the ground water?

- Identify Alternatives: Is this the best location?

In addition to these, it is necessary to investigate existing features such a slope.

If there is a failure of such as features then it is necessary to investigate the failure and suggest remedial works.

The knowledge of subsoil conditions at a site is a prerequisite for a safe and economical design of substructure elements. The field and laboratory studies are carried out for obtaining the necessary information about the surface and subsurface features of the proposed area including the position of the ground water table, these are termed as soil exploration or site investigation.

1.2 Methods of Exploration

Methods of Site Exploration

General: The various types of site investigations are:

- Open excavation.

- Boring.

- Subsurface Sounding.

- Geophysical Methods.

These available methods of exploration can be broadly classified into two categories:

- Direct methods.

- Indirect methods.

The direct method of soil exploration usually consists of sinking a borehole at a predetermined location to the required depth by a method suitable for the site and to obtain fairly intact samples of soils from every stratum encountered or at suitably selected depths.

A samples obtained are utilized to get necessary information about the soil characteristics by means of laboratory tests.

These methods include different sounding and geophysical methods. In sounding methods, the variation in penetration resistance of sample or cone is utilized to interpret some of the physical properties of the strata.

In geophysical methods, the change in subsoil strata are identified by measuring certain physical characteristics, example. Electrical conductance, wave velocity of subsurface deposits. In addition to these methods, probes and aerial photographs, projectiles are also useful in interpreting the soil characteristics.

Geophysical Exploration
General Overview

It is used with advantage to locate boundaries between different elements of the subsoil.

These procedures are based on the fact that the gravitational, electrical, magnetic, radioactive or elastic properties of the different elements of the subsoil may be different.

Differences in the gravitational, magnetic and radioactive properties of deposits near the surface of the earth are seldom large enough to permit the use of these properties in exploration work for civil projects.

The resistivity method based on the electrical properties and the seismic refraction method based on the elastic properties of the deposits have been used widely in maximum civil engineering projects.

Different Methods of Geophysical Explorations

Electrical Resistivity Method

It is based on the various electrical conductivity or the electrical resistivity of different soils.

Resistivity is defined as resistance in ohms between the opposite phases of a unit cube of a material,

$$\rho = \left(\frac{RA}{L} \right)$$

Where ρ is resistivity in ohm-cm:

- R - is resistance in ohms.

- A - is the cross sectional area (cm^2).

- L - is length of the conductor (cm).

The resistivity values of the different soils are listed in table:

Material	Resistivity (Ω -cm)
Massive rock	> 400
Shale and clay	1.0
Seawater	0.3
Wet to moist clayey soils	1.5 - 3.0

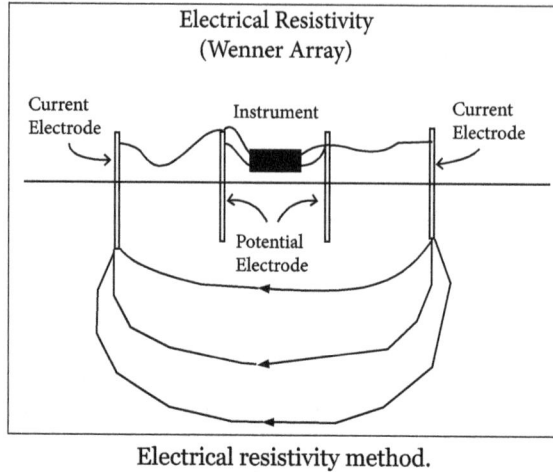

Electrical resistivity method.

Procedure

The set up for the test is given below the figure. In this method, the electrodes are driven approximately 20cms in to the ground and a dc or a very low frequency ac current of known magnitude is passed between the outer electrodes, thereby producing within the soil an electrical field and the boundary conditions. The electrical potential at point C is V_c and at point D is V_d which is measured by means of the inner electrodes respectively,

$$V_c = \frac{1\rho}{2\Pi}\left(\frac{1}{r_1} - \frac{1}{r_2}\right)$$

$$V_D = \frac{\ddot{u}\rho}{\ddot{u}\Pi}\left(\frac{}{{}_3} - \frac{}{{}_4}\right)$$

Where,

ρ - is resistivity.

I - is current.

r_1, r_2, r_3 and r_4 are the distances between the various electrodes as shown in figure.

Potential difference between C and D = V_{CD},

$$\rho = V_c - V_D = \frac{1\rho}{2\Pi}\left[\left(\frac{1}{r_1} - \frac{1}{r_2}\right) - \left(\frac{1}{r_3} - \frac{1}{r_4}\right)\right]$$

$$= \frac{2\Pi V_{CD}}{I}\left[\frac{1}{\left(\frac{1}{r_1} - \frac{1}{r_2}\right) - \left(\frac{1}{r_3} - \frac{1}{r_4}\right)}\right]$$

If $r_1 = r_4 = (r_2/2) = (r_3/2)$ then resistivity is given as,

$$\rho = \frac{2\Pi R r_1}{I}$$

Where,

Resistance $R = (V_{CD}/I)$.

Thus, the apparent resistivity of the soil to a depth approximately equal to the spacing r1 of the electrode can be computed. The resistivity unit is often so designed that the apparent resistivity can be read directly on the potentiometer.

In "transverse profiling" or "resistivity mapping" the electrodes are moved from one place to other place without changing their spacing and the apparent resistivity and any anomalies within a depth equal to the spacing of the electrodes can thereby be determined for a number of points.

Seismic Refraction Method

This method is based on the fact that seismic waves have different velocities in different kinds of soils and besides the wave refract when they cross boundaries between different types of soils.

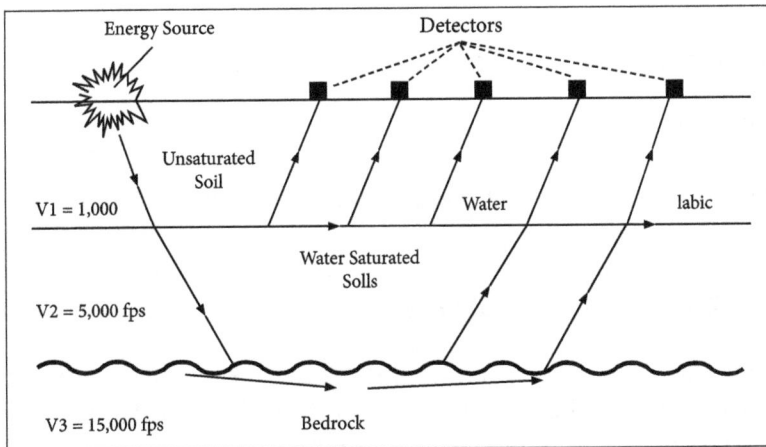

Seismic refraction method.

In this method, an artificial impulse are produced either by detonation of explosive or mechanical blow with a heavy hammer at ground surface or at the shallow depth within a hole. These shocks generate three types of waves:

- Longitudinal or compressive wave or primary (p) wave.

- Transverse or shear waves or secondary (s) wave.

- Surface waves.

It is primarily the velocity of longitudinal or the compression waves which is utilized in this method. The equation for the velocity of the p-waves (V_c) and s-waves (V_s)is given as,

$$V_c = \sqrt{\frac{E(1-\mu)}{(1-\mu)(1-2\mu)\rho}}$$

$$V_s = \sqrt{\frac{E}{2\rho(1+\mu)}}$$

Where,

E - is the dynamic modulus of the soil.

μ - is the Poisson's ratio.

ρ - is density.

G - is the dynamic shear modulus.

These waves are classified as direct, reflected and refracted waves. The direct wave travel in approximately straight line from the source of impulse.

The reflected and refracted wave undergoes a change in direction when they encounter a boundary separating media of different seismic velocities. This method is more suited to the shallow explorations for civil engineering purpose.

The time required for the impulse to travel from the shot point to various points on the ground surface is determined by means of geophones which transform the vibrations into electrical currents and transmit them to a recording unit or oscillograph, equipped with a timing mechanism.

Planning for Subsurface Exploration

The planning of the site exploration program involves location and depth of borings, test pits or other methods to be used and methods of sampling and tests to be carried out. The purpose of the exploration program is to determine, within the practical limits, the stratification and engineering properties of the soils underlying the site.

The principal properties of interest will be the strength, deformation and hydraulic characteristics. The program should be planned so that the maximum amount of information can be obtained at minimum cost. In the earlier stages of an investigation, the information available is often inadequate to allow a firm and detailed plan to be made.

The investigation is therefore performed in the following phases:

- Fact finding and geological survey.

- Reconnaissance.

1. Fact Finding and Geological Survey: The information on dimensions, column spacing, kinds and use of structure, basement requirements and any special architectural considerations of the proposed building. Foundation regulations in the local building code should be consulted for any special requirements. For the bridges the soil engineer should have access to type and span lengths as well as pier loadings. This information will indicate any settlement limitations and can be used to estimate foundation of loads.

2. Reconnaissance: This may be in the form of a field trip to the site which can reveal information on the type and behavior of adjacent sites and structures such as cracks, noticeable sags and possibly sticking doors and windows. The some kinds of local existing structure may influence, to a considerable extent, the exploration program and the best foundation type for the proposed adjacent structure. After the nearby existing structures must be maintained, excavations or vibrations will have to be carefully controlled. Erosion in existing cuts may also be observed.

For highways, run off patterns, along with soil stratification to the depth of the erosion cut, may be observed. Rock outcrops may give an indication of the presence or the depth of bedrock.

1.3   Auguring and Boring

The boring methods are used for exploration at greater depths where direct methods fail. These provide each disturbed as well as undisturbed samples depending upon the method of boring.

Auger Boring

This method is fast and economical, using simple light, flexible and inexpensive instruments for maximum to small holes. It is very suitable for soft to stiff cohesive soils and also can be used to determine ground water.

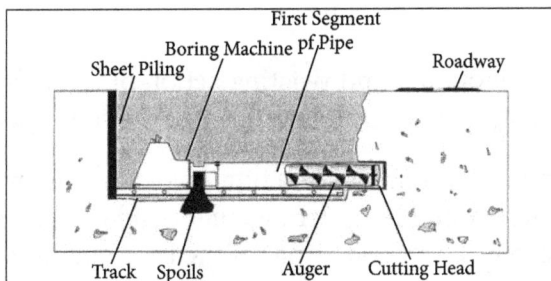

Augers.

Soil removed by this is disturbed but it is better than wash boring, percussion or rotary drilling. It is not suitable for very hard or cemented soils, very soft soils.

1.4   Wash Boring and Rotary Drilling

Wash Boring

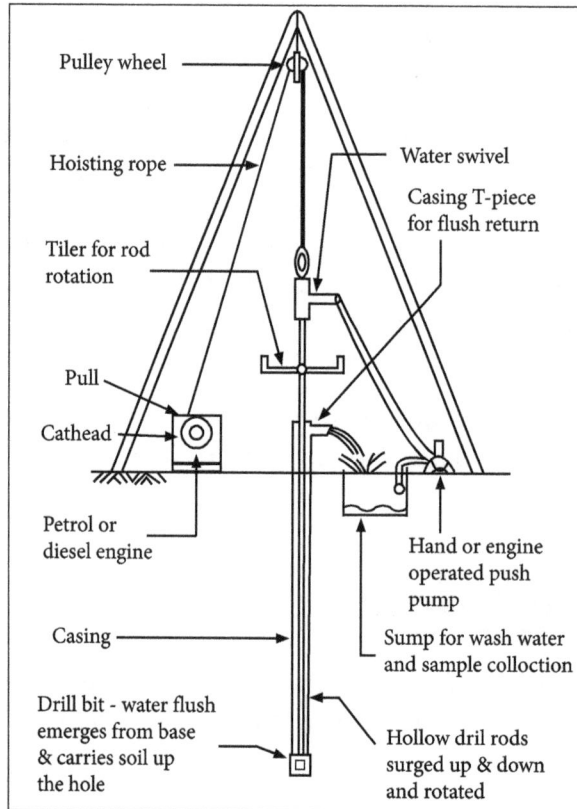

Wash Boring.

It is a popular method due to the use of limited equipments. The benefits and use of inexpensive and easily portable handling and drilling equipments. Here first an open hole is formed on the ground so that the soil sampling or rock drilling operation can be done below the hole.

The hole is advanced by chopping and twisting action of the light bit. Cutting is done by forced water and water jet under pressure through the rods operated inside the hole.

In India the "Dheki" operation is used, i.e., a pipe of 5cm diameter is held vertically and filled with water using horizontal lever arrangement and by the process of suction and application of pressure, soil slurry comes out of the tube and pipe goes down. This can be done up to a depth of 8m –10m.

The change of colour of soil coming out with the change of soil character can be identified by any experienced person. It gives an completely disturbed sample and is not suitable for very soft soil, fine to medium grained cohesion less soil and in cemented soil.

Rotary Drilling

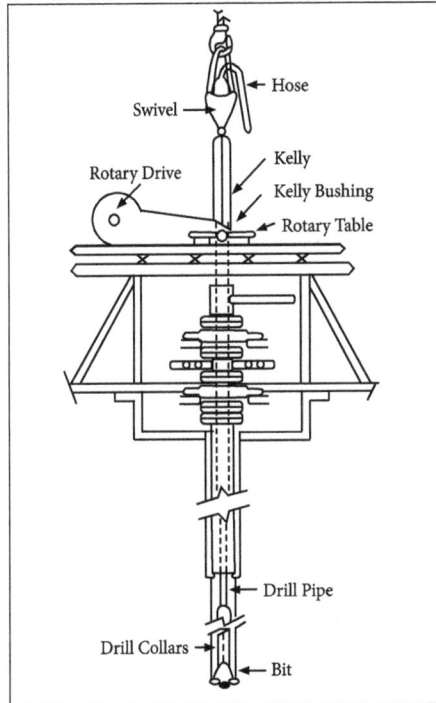

Rotary drilling system.

The rotary drilling method of boring is useful in case of highly resistant strata. It is related to find out the rock strata and also to access the quality of rocks from cracks, fissures and joints.

It can conveniently be used in sands and silts also. There are bore holes are advanced in depth by rotary percussion method which is similar to wash boring technique.

A heavy string of the drill rod is used for choking action. The broken rock or soil fragments are removed by circulating water or drilling mud pumped through the drill rods and bit up through the bore hole from which it is collected in a settling tank for recirculation.

If the depth is medium and the soil stable, water alone can be used. Although, drilling fluids are useful as they serve to stabilize the bore hole.

The drilling fluid causes stabilizing effect to the bore hole partly due to higher specific gravity as compared with water and partly due to formation of mud cake on the sides of the hole.

As the stabilizing effect is imparted by these drilling fluids no casing is required if drilling fluid is used. This method is suitable for boring holes of diameter 10cm or more then preferably 15 to 20cm in most of the rocks. It is uneconomical for holes less than 10cm diameter. The depth of different strata can be detected by inspection of cuttings.

1.5   Depth of Boring

The depth of soil affected by the load transmitted by the foundation determines the required depth of boring in the all process.

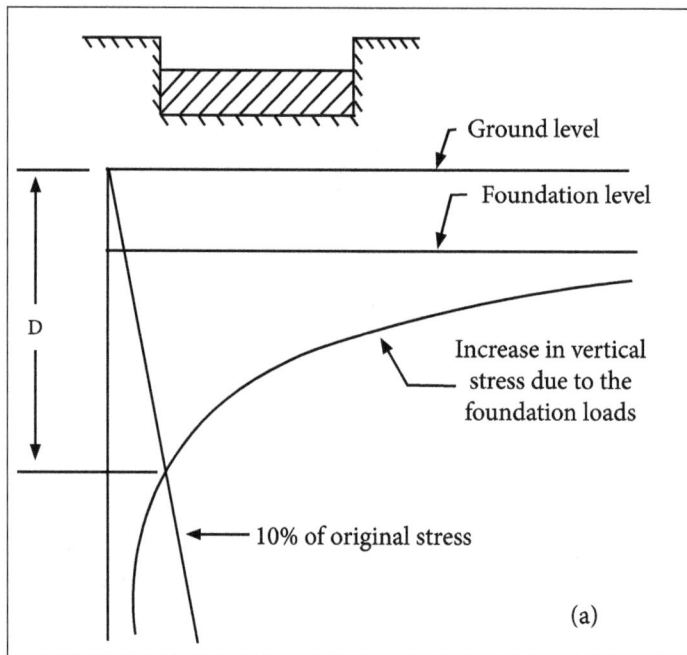

Building.

The additional stresses imposed by the foundation to a depth of one and half times the total width of foundation is still found to be about 20% minimum the foundation base. At the depth equal to two times the width of foundation, this stress is about 10%.

Considering the effect of stress felt, the boring should be always taken to a depth at least equal to one and half times of the foundation width below the base of the foundation.

The depth of boreholes to be made under different foundation conditions are shown in figure. Some additional information regarding the depth of boring can be obtained from the table given below:

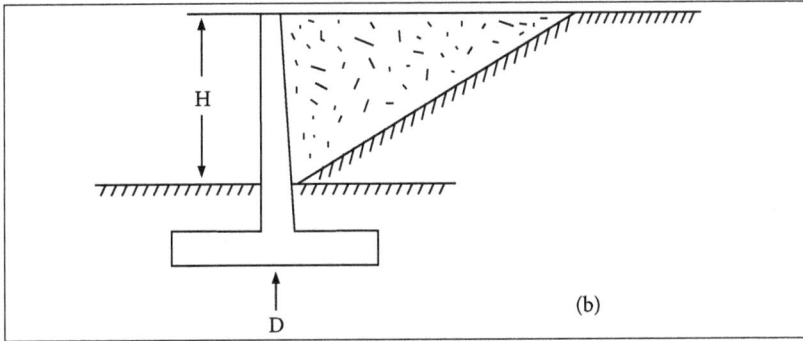

Retaining wall.

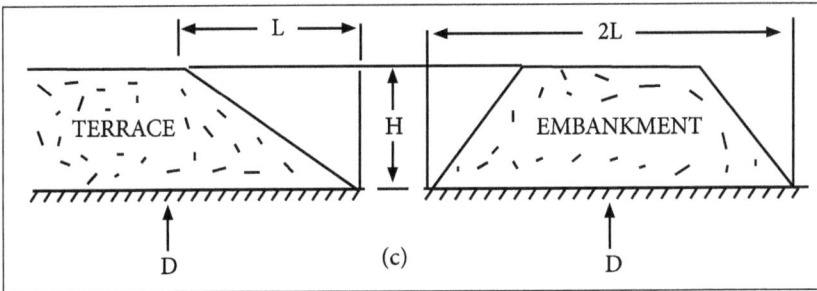

Terraces and fill.

Table Depth of Boring According as Structure Type:

Structure type	Design Consideration	Depth of Boring	Remark
Building	Settlement, total and differential	10 m minimum or up to depth where the increase in stress due to structure is 10% of stress imposed by structure	D = Depth of exploration.
Retaining walls	Bearing capacity and settlement	D = 0.5 to 2H	H = height of wall D = Depth of hole and
Terraces and fill	Bearing capacity and settlement	D = 1.25 L for terraces and 0.5 L for fill	D = Depth of exploration and L = Projected slope length
Depth cuts	Stability of slopes	D = 0.75 B to 1.0 B	D = Depth of exploration and B = Bottom width of cut
Earth dam and levee	Bearing capacity	D = L	D = Depth of exploration and B = Bottom width of dam
Concrete dam	Bearing capacity	D = 1.5 to 2H	D = Depth of exploration and H = Height of dam
Highways, railways and airports	General stability and drainage conditions	1 to 2 m for light loads and 2 to 3.5 m for heavy loads	
Tunnels	Stability of materials and pressure exerted against tunnel lining	D = B	D = Depth of exploration and B = Gross width of tunnel

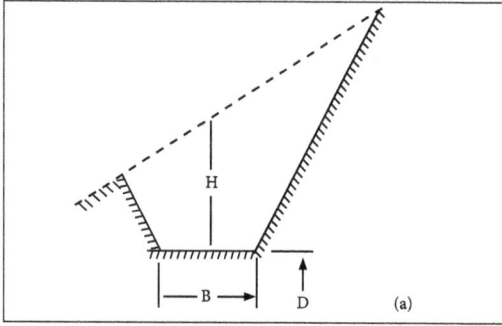

Depth cuts.

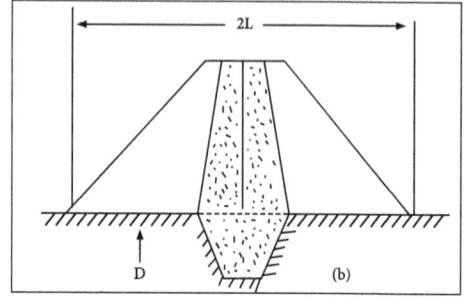

Earth dam and levee.

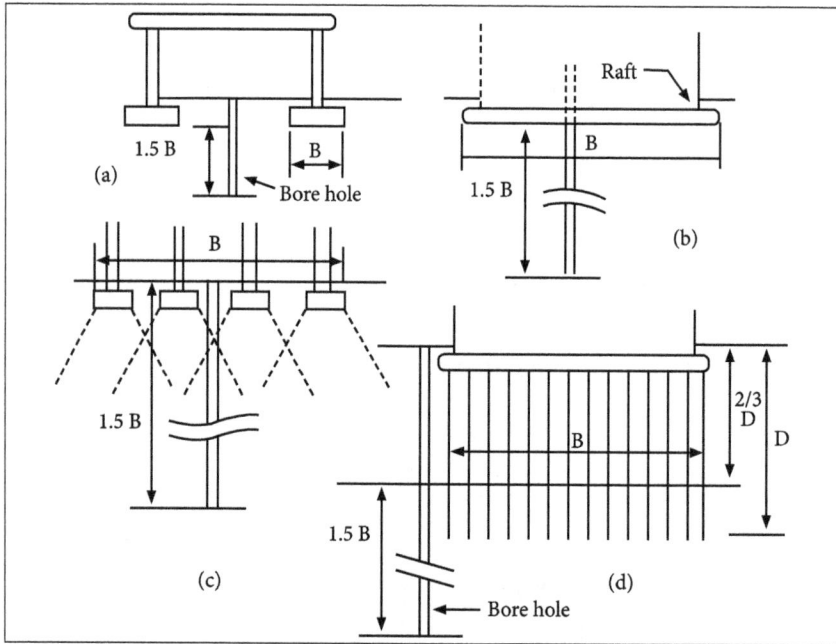

Depth of Boring.

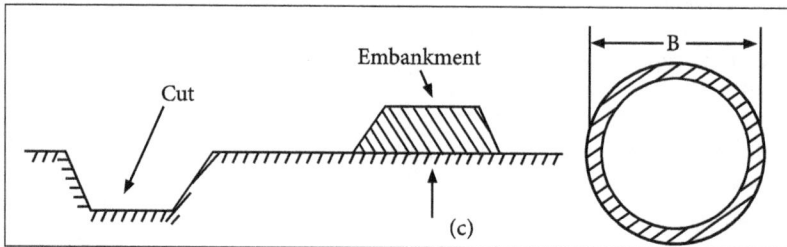

Concrete dam.

Soil Exploration Report

At the end of the soil exploration program, the soil and rock samples, collected from the field are subjected to visual observation and laboratory tests. Then, a soil exploration report is ready to be used by the planning and design office.

Any soil exploration report ought to contain the following information:

- Scope of investigation.
- General description of the proposed structure for that the exploration has been conducted.
- Geological conditions of the site.
- Details of boring.
- Drainage facilities at the site.
- Description of subsoil conditions as determined from the soil and rock samples collected.
- Any anticipated construction problems.
- Details of foundation recommendations and alternatives.
- Ground water table as discovered from the boreholes.
- Limitations of the investigation.

The following graphic presentations additionally required to be attached to the soil exploration report:

- Site location map.
- Location of borings with respect to the proposed structure.
- Boring logs.
- Laboratory test results.
- Other special presentations.
- The boring log is the graphic representation of the details gathered from every borehole.

1.6   Spacing of Bore Hole

The location and spacing of trial pits or bore holes for a particular site require special consideration:

- The pits or the bore holes should be so located so as to give adequate information in respect of changes in properties of the underlying strata with depth.

- A number and spacing of the test pits or bore holes to be adopted for a site will depend upon the area of the plot along with the type of structure to the built.

- For a plot of 0.4 hectare we may have one trial pit or bore hole near both corner and one in the middle. For smaller size of plot and for ordinary structure, one trial pit or bore hole near the center should be sufficient.

- The depth of the pit or bore hole will depend upon the characteristics of the soil along with this type of structure and shape, size and loading condition.

- As a thumb rule, its depth should can be one and a half times the probable width of the footing or 1.5 m whichever is more.

- In case of weak soils, although the test pits or the bore holes should be taken to a depth at which the loads can be carried by the soil without undesirable settlement.

Table presented below provides some guideline as to the minimum numbers and spacing of boreholes:

Types of job	Types of soil in horizontal extent spacing (m)			Minimum no. of boring
	Uniform	Average	Erratic	
1 or 2 storied structure	60	30	15	3
Multi-storied structure	45	30	15	4
Bridge piers, abutments	–	30	15	1-2 per foundation
Transmission tower	–	30	15	1-2 per foundation
Highway and airports	300	150	100	–
Borrow pits	300-150	150-100	30-15	–

The depth of the pit or bore hole will depend upon the characteristics of the soil as well as the type of structure, its shape, size and loading condition.

As a thumb rule, its depth should be one and a half times the probable width of the footing or 1.5 m whichever is more.

In case of weak soils, although the test pits or the bore holes should be taken to a depth at which the loads can be carried by the soil without undesirable settlement.

Table presented below provides some guideline as to the minimum numbers and spacing of boreholes.

1.7 Sampling Techniques

Soil samples can be of two types:

- Disturbed samples.

- Undisturbed samples.

A disturbed sample is that in which the natural structure of soil gets partially or fully modified and destroyed although with suitable precautions the natural water content may be preserved.

Such as a soil sample should, although, be representative of the natural soil by maintaining the original proportion of the various particles intact.

An undisturbed sample is that in which the natural structure and properties remain preserved. The sample disturbance depends upon the design of the samplers and the method of sampling. To take undisturbed samples from bore holes properly designed sampling tools are required.

The sampling tube when forced into the ground should cause as little re-moulding and disturbance as possible.

The design features of the sampler that govern the degree of disturbance are:

- Cutting edge.

- Inside wall friction.

- Non-return valve.

The figure shows given below a typical cutting edge of a sampler, with the lower end of the sampler, with the lower end of the sampler tube. The following terms are defined with respect to the diameters marked in figure,

$$\text{Area ratio} = \frac{D_2^2 - D_1^2}{D_1} \times 100$$

$$\text{Inside clearance} = \frac{D_3 - D_1}{D_1} \times 100$$

$$\text{Outside clearance} = \frac{D_2 - D_4}{D_4} \times 100$$

The area ratio should be as low as possible. It should not be greater than 25 percent. For soft sensitive soil, it should preferably not exceed to percent.

The inside clearance should lie between 1 to 3 % and the outside clearance should not be much greater than the inside clearance. The walls of the sampler should be smooth and should be kept properly oiled so that wall friction is minimum.

Lower value of inside clearance allows the elastic expansion of soil and reduces the frictional drag. The non-return valve, invariably provided in samplers, should permit easy and quick escape of water and air when driving the sampler.

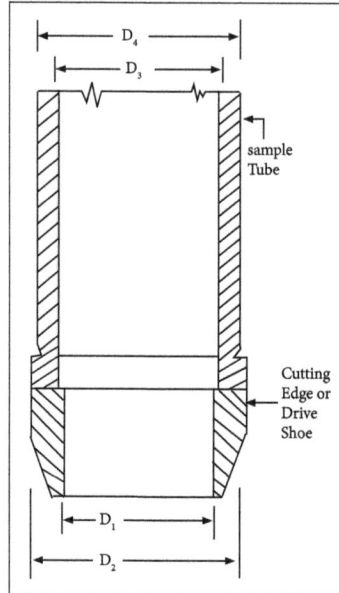

Lower end of a sampler.

Types of Samplers

The samplers are classified to thick wall or thin wall samplers depending upon the area ratio. Thick wall samplers are those having the area ratio greater than 10%.

Depending upon the mode of operation, samplers may be classified in the following three.

Common types:

- Open drive sampler.
- Rotary sampler.
- Stationary piston sampler.

The open drive sampler is a tube open at its lower end. The sampler head is provided with vents to permit water and air to escape during driving.

The check valve helps to retain sample when the sampler is lifted up. The tube may be seamless or it may be split in two parts. In the latter case it is known as split spoon sampler.

Rotatory samplers are the core barrel type having an outer tube provided with cutting teeth and a removable thin wall liner inside. It is used for firm to hard cohesive soils and cemented soils.

The stationary piston sampler consists of a sample cylinder and the piston system. During lowering of the sampler through the hole, the lower end of the sampler is kept closed with the piston.

When the desired sampling elevation is reached, the piston rod is clamped thereby keeping the piston stationary and the sampler tube is advanced down into the soil.

The sampler is then lifted up, with piston rod clamped in position. The sampler is more suitable for sampling soft soils saturated sands.

1.8   Representative and Undisturbed Sampling

Disturbed Samples

The structure of the soil is disturbed to the considerable degree by the action of the boring tools or the excavation equipments.

The disturbances can be classified in following basic types:

- Change in the stress condition.
- Change in the water content and the void ratio.
- Disturbance of the soil structure.
- Chemical changes.
- Mixing and segregation of soil constituents.

The causes of the disturbances are listed below:

- Method of advancing the borehole.
- Mechanism used to advance the sampler.
- Dimension and type of sampler.
- Procedure followed in sampling and boring.

Undisturbed Samples

It retains as closely as practicable the true in situ structure and water content of the soil. For undisturbed sample the stress changes cannot be avoided. The following requirements are looked for:

- No change due to disturbance of the soil structure.
- No change in void ratio and water content.
- No change in constituents and chemical properties.

Requirement of Good Sampling Process

Inside clearance ratio,

$$(C_i) = \frac{D_s - D_c}{D_c} \times 100 \ \%$$

The soil is under great stress as it enters the sampler and has a tendency to laterally expand.

The inside clearance should be large enough to allow a part of lateral expansion to take place, but it should not be so large that it permits excessive deformations and causes disturbances of the sample.

For a good sampling process, the inside clearance ratio should be within 0.5 to 3 %. For sands silts and clays, the ratio should be 0.5 % and for stiff and hard clays, it should be 1.5 %. For stiff expansive type of clays, it should be 3.0 %.

Area ratio,

$$(A_\gamma) - \frac{D_v^{\,2} - D_c^{\,2}}{D_c^{\,2}} \pi \ 100\%$$

Recovery ratio,

$$(R) = \frac{L}{H} \times 100 \ \%$$

Where,

L - is the length of the sample within the tube.

H - is the depth of penetration of the sampling tube.

It represents the disturbance of the soil sample. For good sampling the recovery ratio should be 96 to 98 %.

Wall friction can be reduced by suitable inside clearance, smooth finish and oiling. The non-returned wall should have large orifice to allow air and water to escape.

Stabilization of Boreholes: Typical Bore Log

Stabilization of borehole is done to stop the boreholes from caving in. Uncased borehole are usually stable once they are at shallow depth and are well above the ground water, table. With the increase of depth and the existence of ground water, the risk of caving increases rapidly. Therefore borehole needs stabilization. Various methods of stabilization of boreholes are as follows:

1. Stabilization with Water: In this method, the boreholes are kept filled with water. The water is then circulated to eliminate cutting soils from the bottom of the borehole. The water exerts downward and lateral thrust that counteracts soil and pore water pressures. It helps stopping caving in of walls and heaving of bottom. Water stabilization will be used in rock and stiff cohesive soils.

2. Stabilization with Drilling Fluid: Borehole will often be stabilized by filling it with a suitably proportioned drilling fluid or mud. Some products such as Bentonite or aqua-gel are usually applied as drilling fluid. This material forms an impervious lining on the walls of the borehole and helps preventing caving in of walls.

3. Stabilization with Casing: The simplest method of protecting the borehole from caving in, is casing the borehole with steel pipes. The casing pipe is generally driven by repeated blows of the hammer. The lower end of the casing pipe is usually protected by a casing shoe.

4. Stabilization with Grouting: A borehole can be stabilized by cement grout for a difficult region in rock such as cavities and faults.

1.9 Methods: Split Spoon Sampler, Thin Wall Sampler and Stationery Piston Sampler

Standard Split Spoon Sampler

Standard split spoon sampler consists of tool-steel driving shoe at the bottom, a steel tube in the middle and a coupling at the top. The steel tube in the middle has inside and outside diameters of 34.9mm and 50.8mm, respectively.

When a bore hole is advanced to a desired depth, the drilling tools are removed. The split-spoon sampler is attached to the drilling rod and then lowered to bottom of the bore hole.

The sampler is driven in to the soil at the bottom of the bore hole by means of hammer blows. The hammer blows occur at the top of the drilling rod.

The hammer weights 623N. For both blow, the hammer drops a distance of 0.762 m. The number of blows required for driving the sampler through the three 152.4 mm interval is recorded. The sum of the number of blows required for driving the last two 152.4mm intervals is referred to as the standard penetration number N. It is also generally called as blow count.

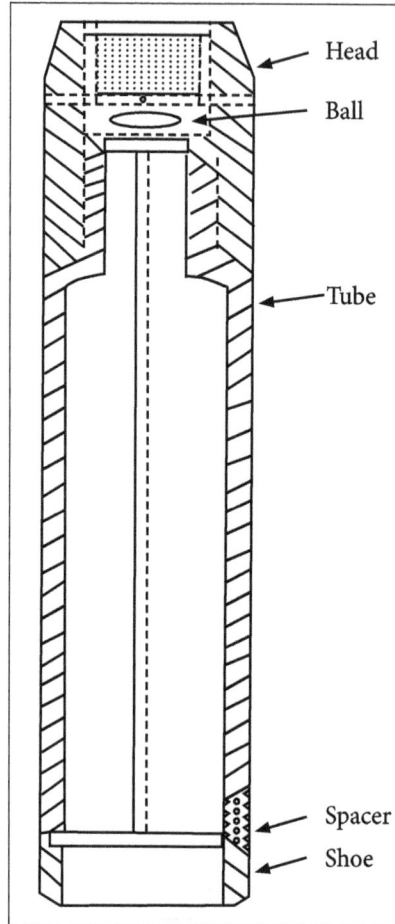

Split spoon sampler.

After driving is completed, the sampler is withdrawn and the shoe and coupling are removed. The soil sample collected inside the split tube is then removed and transferred to the laboratory in small glass jars. Determination of the standard penetration number and collection of split-spoon samples are usually done at 1.5m.

It consists of a longitudinally split tube or barrel fitted with a shoe and a sampler head with provision for air release. The splitting aspect of the sampler permits it to be opened for a sample examination and for onward transmission in sample containers to laboratories. Samples obtained using this sampler is rated as representative. This sampler is suited for sand and is used only in the standard penetration test.

Split-soon samplers may be provided with a liner, which is a thin metal or plastic tube fitted within the split spoon, in which case it is called a composite sampler. The purpose of the liner is to protect the sampler during shipping, handling and storage.

Thin-Walled Sampler

To obtain undisturbed samples in soft to firm clays and plastic silts, thin-walled samplers may be used (IS: 11594, 1985).

No separate cutting shoe is attached to the lower end, but the sampler's lower end is itself machined to serve as a cutting edge. The good-quality undisturbed samples are possible if A < 10% and the soil is not disturbed during the boring operation. This sampler can be used more conveniently in trial pits and shallow boreholes.

Open-Drive and Piston Samplers

Undisturbed samples may be obtained from boreholes by open drive samplers or piston samplers. Open drive samples consist of thin-walled tubes which are pushed or driven in to the soil at the bottom of the hole and then rotated to detach the lower end of the sample from the soil. Most soft or moderately stiff cohesive soil can be sampled without extensive disturbance in thin-walled seamless steel tubes having diameter is not less 50 mm.

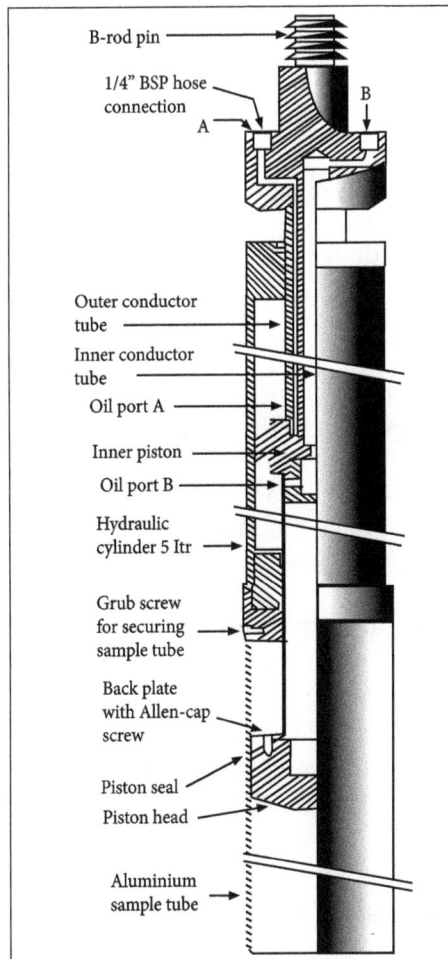

Piston type Sampler.

The lower end of the tube is sharpened to from a cutting edge and the other end is machined for attachment to the drill rods. The entire tube is pushed or driven into the soil at the bottom of the hole and is removed from the sample. The two ends of the tube are then sealed and the sample is shifted to the laboratory.

The good quality undisturbed samples are obtained from piston samplers which use thin-walled sampling tubes with a piston inside. While the tube is being lowered to the bottom of the drill hole, the piston rods and piston are held at the bottom of the sampler by means of a drill rod which rises to the top of the bore hole. The presence of the piston prevents excess soil from squeezing into the tube and thus, maintains the integrity of the sample.

1.10 Penetration Tests

Field Testing

There are many different kinds of tests that can be performed at the time of drilling or project site. The three kinds of field tests are most generally used geotechnical practice:

- Standard penetration test (SPT).
- Static cone penetration test (CPT).
- Dynamic cone penetration test (DCPT).

Penetration Tests

These tests involve the measurement of the resistance to penetration of a sampling spoon, a cone or other shaped tools under dynamic or static loadings.

The resistance is empirically correlated with some of the engineering properties of soil such a density index, bearing capacity, consistency, etc. The values of these tests lie in the amount of experience behind them.

These tests are useful for common exploration of erratic soil profiles, for finding depth to bed rock or hard stratum and to have an approximate indication of the strength and other properties of soils, particularly the cohesion less soils, from which it is difficult to obtain undisturbed samples.

There are two commonly used tests are the standard penetration test and the cone penetration test.

Standard Penetration Test

The standard penetration test is carried out in a borehole, while the DCPT and SCPT are carried out without a borehole. All the three tests measuring the resistance of the soil strata to penetration by a penetrometer.

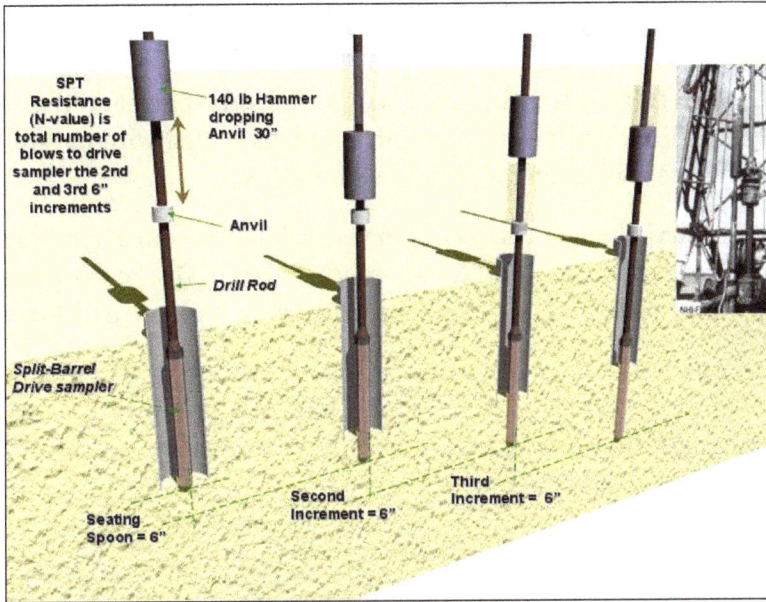

Standard penetration test.

The useful empirical correlations between penetration resistance and soil properties are available for use in foundation design.

This is the most extensively used penetrometer test and employs a split-spoon sampler, which consists of a driving shoe, a split-barrel of circular cross-section which is longitudinally split into two parts and a coupling. IS: 2131-1981 gives the standard for carrying out the test.

Procedure

- The borehole is advanced to the required depth and the bottom cleaned.

- The split-spoon sampler, attached to standard drill rods of required length is lowered into the borehole and rested at the bottom.

- The split-spoon sampler is driven into the soil for a distance of 450mm by blows of a drop hammer of 65 kg falling vertically and freely from a height of 750 mm. The number of blows required to penetrate every 150 mm is recorded while driving the sampler.

- The number of blows required for the last 300 mm of penetration is added together and recorded as the N value at the particular depth of the borehole. The number of blows required to effect the first 150mm of penetration, known the seating drive, is disregarded.

- The split-spoon sampler is then withdrawn and is detached from the drill rods. The split-barrel is disconnected from the cutting shoe and the coupling. The soil sample collected inside the split barrel is carefully collected so as to preserve the natural moisture content and transported to the laboratory for tests.

- Sometimes a thin liner is inserted within the split-barrel so that at the end of the SPT, the liner containing the soil sample is sealed with molten wax at both its ends before it is taken away to the laboratory.

- The SPT is carried out at every 0.75 m vertical intervals in a borehole. This can be increased to 1.50 m if the depth of borehole is large. Due to the presence of boulders or rocks, it may not be possible to drive the sampler to a distance of 450 mm.

- In such case, the N value can be recorded for the first 300 mm penetration. The boring log shows refusal and the test is halted if 50 blows are required for any 150mm penetration 100 blows are required for 300m penetration 10 successive blows produce no advance.

Precautions

The drill rods must be of standard specification and should not be bent condition. The split spoon sampler should be in good condition and the cutting shoe must be free from wear and tear. The drop hammer should be of the right weight and the fall should be free, frictionless and vertical.

The SPT is carried out at every 0.75 m vertical intervals in a borehole. This can be increased to 1.50 m if the depth of borehole is large. Due to the presence of boulders or rocks, it may not be possible to drive the sampler to a distance of 450 mm.

In such a case, the N value can be recorded for the first 300 mm penetration. The boring log shows refusal and the test is halted if 50 blows are required for any 150mm penetration 100 blows are required for 300m penetration 10 successive blows produce no advance. The height of fall should be exactly 750 mm. Any change from this will seriously affect the N value.

The bottom of the borehole should be properly cleaned before the test is carried out. If this is not done, the test gets carried out in the loose, disturbed soil and not in the undisturbed soil.

When a casing is used in borehole, it should be ensured that the casing is driven just short of the level at which the SPT is to be carried out. Otherwise, the test gets carried out in a soil plug enclosed at the bottom of the casing.

When the test is carried out in a sandy soil below the water, it must be ensured that the water level in the borehole is always maintained slightly above the ground water level.

If the water level in the borehole is lower than the ground water level, quick condition may develop in the soil and very low N values may be recorded.

In spite of all these imperfections, SPT is still extensively used because the test is simple and relatively economical. It is the only test that provides representative soil samples both for visual inspection in the field and for natural moisture content and classification tests in the laboratory.

A SPT values obtained in the field for sand have to be corrected before they are used in empirical correlations and design charts. IS: 2131-1981 recommends that the field value of N be corrected for two effects, namely:

- Effect of overburden pressure.

- Effect of dilatancy.

Correction for Overburden Pressure

The many investigators have found that the penetration resistance or the N value in a granular soil is influenced by the overburden pressure of two granular soils possessing the same relative density but having different confining pressures, the one with a higher confining pressure gives a higher N value. After the confining pressure increases with depth, the N values at shallow depths are underestimated and the N values at larger depths are overestimated.

To allow for this, N values recorded from field tests at different effective overburden pressures are corrected to a standard effective overburden pressure.

Static Cone Penetration Test

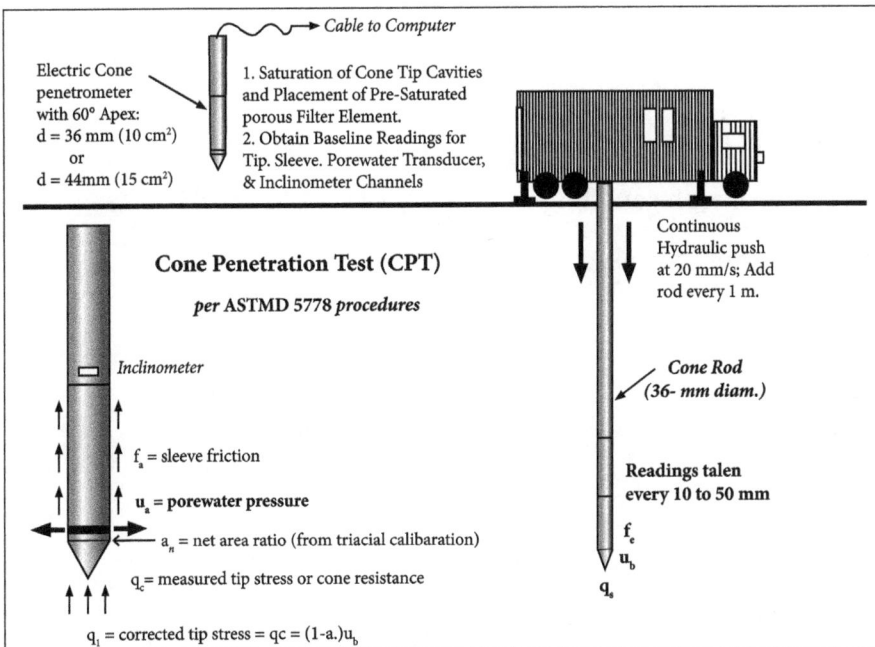

Static cone penetration test.

At field SCPT is widely used of recording variation in the in-situ penetration resistance of soil in cases where in-situ density is disturbed by boring method and SPT is unreliable below water table.

The test is very useful for soft clays, soft silts, medium sands and fine sands.

Procedure

By this test basically by pushing the standard cone at the rate of 10 to 20 mm/sec in to the soil and noting the friction, the strength is determined. After installing the equipment as per IS-4968,the sounding rod is pushed in to the soil and the driving is operated at the steady rate of 10 mm/sec approximately so as to advance the cone only by external loading to the depth for which a cone assembly available.

For finding combine cone friction resistance, the shearing strength of the soil q_s and tip resistance q_c is noted in gauge and added to get the total strength.

Limitations

This test is unsuitable for gravelly soil and soil having SPT N value greater than 50. And also in dense sand anchorage becomes too cumbersome and expensive for such a cases Dynamic SPT can be used. This test is also unsuitable for field operation since erroneous value obtained due to presence of brick bats, loose stones etc.

Correction in SCPT Test

$$C_N = (m + nm_1) \times 10 KN/m^2$$

Here,

m = Mass of cone =1.1 Kg.

m_1 = Mass of each sounding roads = 1.5 Kg.

n = Number of rods used.

Friction ratio,

$$f_r = \frac{q_s}{q_c} \times 100\%$$

f_r = Friction ratio.

q_s = Measured site/slip friction.

q_c = Tip resistance/point resistance.

Then, sensitivity of soil is measured,

$$S_t = \frac{10}{f_r}(f_r \text{ in } \%)$$

Where,

S_t = Sensitivity of soil.

For Cohesive soil (Undrained shear strength),

$$S_u = \frac{q_c - p_o}{N_k}$$

p_o = overburden pressure = γZ

N_k = Cone factor = 15 to 20(depends on the plasticity index of soil)

q_c = 612.6 + 587.5I_c

Her I_c = Consistency index of soil

q_c is measured in KPa

Schmertman

$$\frac{q_{c.ocr}}{q_{c.NC}} = 1 + K\left(\frac{K_{o.ocr}}{K_{o.NC}} - 1\right) \qquad (k = 0.75)$$

$$\frac{K_{o.ocr}}{K_{o.NC}} = (OCR)^n \qquad (n = 0.32 \text{ to } 0.52)$$

Relationship between angle of internal friction and undrained shear strength.

For gravelly silt,

$$\phi = \left(29 + \sqrt{q_c}\right) + 5^\circ$$

For silty sand,

$$\phi = \left(29 + \sqrt{q_c}\right) - 5^\circ$$

SPT & SCPT relation,

$$q_c = K.N$$

Sl. No	Soil type	$K.\left(\dfrac{q_c}{N_{60}}\right)$
1	Silt, sand silt and slightly cohesive sand silt mix	0.1 to 0.2
2	Clean fine sand to medium sand and slightly silty sand	0.3 to 0.4
3	Coarse sand and sand with gravel	0.5 to 0.7
4	Sandy gravel and gravel	0.8 to 1.0

Geophysical Tests

Geophysical surveys are used in one of two roles. Firstly, to aid a rapid and economical choice between a number of alternative sites for a proposed project, prior to the design investigation and secondly, as part of the detailed site assessment at the chosen location.

Geophysical methods also have a important role to play in the resource assessment and the determination of engineering parameters. The recently issued British Code of Practice for Site Investigations sets out four primary applications for engineering geophysical methods.

Geological Investigations

The role to play in mapping stratigraphy, determining the thickness of superficial deposits and the depth to engineering rock head, establishing weathering profiles and the study of particular erosional and structural features:

- Resources assessment: Location of aquifers and determination of water quality; exploration of sand and gravel deposits and rock for aggregate; identification of clay deposits.

- Determination of engineering parameters: Such as dynamic elastic moduli needed to solve many soil-structure interaction problems; soil coercivity for a pipeline protection studies; rock rip ability and rock quality.

- Detection of voids and buried artifacts: Examples: natural cavities, mineshafts, old foundations, wrecks at sea, pipelines, etc.

1.11 Bore Log Report

A detailed record of boring operations and other tests carried out in the field is an essential part of the field work. The bore hole log is made during the boring operation.

The soil is classified based on the visual examination of the disturbed samples collected. A typical example of a bore hole log is obtained in the following table.

Table:

Job No Date:
Project: BH NO: Location: GL: Boring Method WTL: Dia of BH: Supervisor:

Soil Type	Level m	Depth m	SPT				Sample type	Remark
			15 cm	15 cm	15 cm	N		
Yellowish Stiff clay								
Greyish sandy silt medium dense								
Greyish silt sand dense								
Blackish very stiff clay								

1.12 Data Interpretation

Interpretation from Test Results

From the test results, a load-settlement curve should be plotted. The allowable pressure of the prototype foundation for an assumed settlement may be found and by making use of the following equations as suggested by Terzaghi and Peck.

For granular soils,

$$S_f = S_F \left(\frac{B(b_F + 0.3)}{b_F(B + 0.3)} \right)^2$$

For clay soils,

$$S_f = S_F \frac{B}{b_F}$$

Where,

S_f = Permissible settlement of the foundation in mm,

S_p = Settlement of the plate in mm,

B = Size of plate in meters,

b_p = Size of plate in meters.

The permissible settlement S_f for a prototype foundation should be known. Normally a settlement of 2.5 cm is recommended. In the equation the values of S_f and b_p are known.

The unknowns are S_p and B. The value of S_p for any assumed value of B may be found out from the equation.

Using the plate load settlement curve the value of the bearing pressure corresponding to the computed value of S_p is found out. This bearing pressure is the safe bearing pressure for a given permissible settlement S_f.

1.13 Strength Parameters and Liquefaction Potential

The liquefaction occurs in saturated soils. All pores are completely filled with water. The water in the pores exerts the pressure on soil particles known as Pore Pressure that influences how tightly the soil particles are pressed together.

Occurrence of Liquefaction

Prior to an earthquake: the pore pressure is low earthquake shaking causes pore pressure to Increase: Undrained Condition.

Pore pressure = Overburden press/Confining stress Effective stress is Zero. No Shear Strength.

Soil particles starts readily move with respect to both other due to zero shear strength.

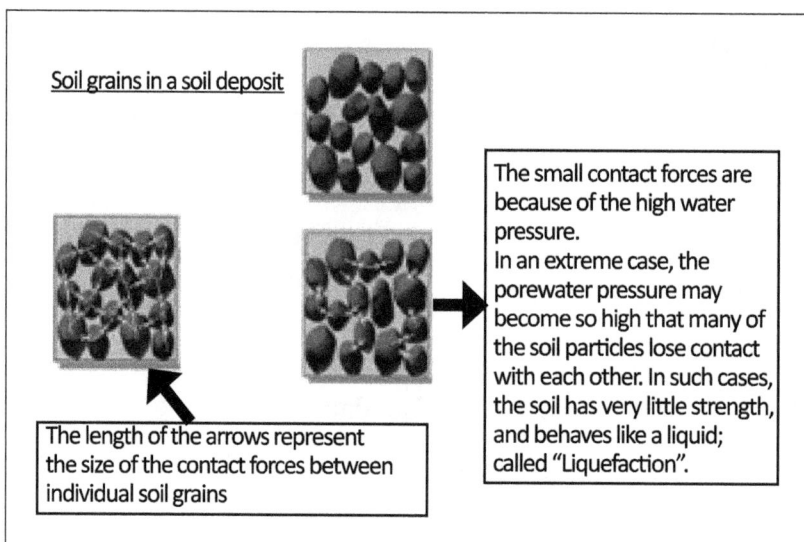

Soil grains in a soil deposit

The small contact forces are because of the high water pressure.
In an extreme case, the porewater pressure may become so high that many of the soil particles lose contact with each other. In such cases, the soil has very little strength, and behaves like a liquid; called "Liquefaction".

The length of the arrows represent the size of the contact forces between individual soil grains

Mechanism of liquefaction.

1.14 Selection of Foundation based on Soil Condition

Soil type affects selection of types of foundations, foundation depth and foundation sizes.

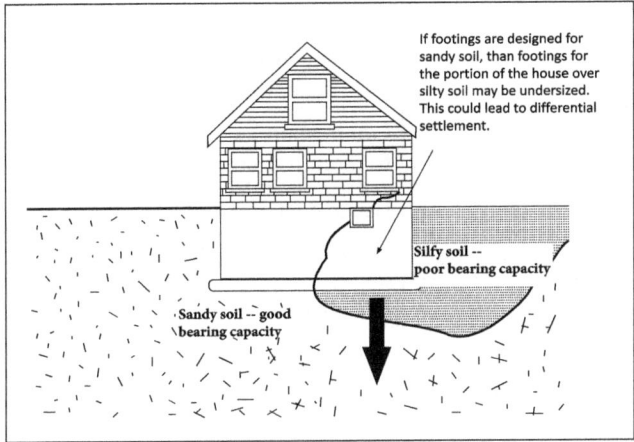

Soil condition.

Following are the considerations for foundation based on soil types:

Soil Type	Foundation		Trouble areas
	Types	Reason for use	
Sand	Footings	Easy to construct and economical.	Bearing capacity may be a problem but in most cases it is sufficient. Excessive settlement in wet and loose deposits. Confining pressure is usually low.
	Retaining Structures	Must be used since sand cannot support themselves.	
	Deep foundations (Piles)	Uses friction resistance but low in bearing capacity.	
Clay	Footings	Economic but may have problem with bearing capacity in saturated clays.	Low bearing capacity. Generally low shear strength when wet. High consolidation in soft clays Swelling is possible. Over-consolidated clays may contain cracks and fissures.
	Retaining Structures	Clays are self-supportive up to a certain height (critical). Must be used if height increases beyond the critical.	
	Deep foundations (Piles)	If bearing capacity is low, piles may be driven to rock. May change formation of clay.	

Following table shows solutions to some problems in foundations based on soil types:

Soil Type	Nature of problem	Possible solution
Sand	Settlement	Loose sands must be compacted
		Lowering water table may result in sand densification
	Bearing Capacity	Compaction increases cohesion and friction thus bearing capacity increases
		Use of deep foundation
Clay	Consolidation	Lowering water table
		Pre-loading
		Drive pile to rock
	Bearing capacity	Compaction
		Use of deep foundations
	Expansion or swelling	Treat or stabilize soil
		Maintain constant water table
		Alter soil nature (similar to stabilization)
		Include swell pressure in design

2

Shallow Foundation

2.1 Introduction to Foundation

Foundations may be broadly classified as:

- Shallow foundations.

- Deep foundations.

According to Terzaghi, a foundation is shallow if its depth is equal to or less than its width. In the case of deep foundations, the depth is equal to or greater than the width. Apart from deep strip, rectangular or square foundations, other common forms of deep foundations are:

- Pier foundation.

- Pile foundation.

- Well foundation.

2.2 Location and Depth of Foundation

There are many factors affects the determination of depth of foundation. Calculation for foundation depth is done based on some kind of soil, loads from structure, ground water table, bearing capacity of soil and other factors.

General factors to be considered for determining depth of foundation are:

- Load applied from structure to the foundation.

- Depth of water level below the ground surface.

- Bearing capacity of soil.

- Kinds of soil and depth of layers in case of layered soil.

- Depth of adjacent foundation.

The minimum depth of foundation should be considered to ensure that the soil is having the required safe bearing capacity as assumed in the design. It is advised to carry out soil investigation before deciding on the depth of foundation.

The soil investigation report will suggest the foundation depth based on the soil properties, kind of structure, depth of water table and all other variable that should be considered. This report provides bearing capacity of soil at variable levels and at different locations.

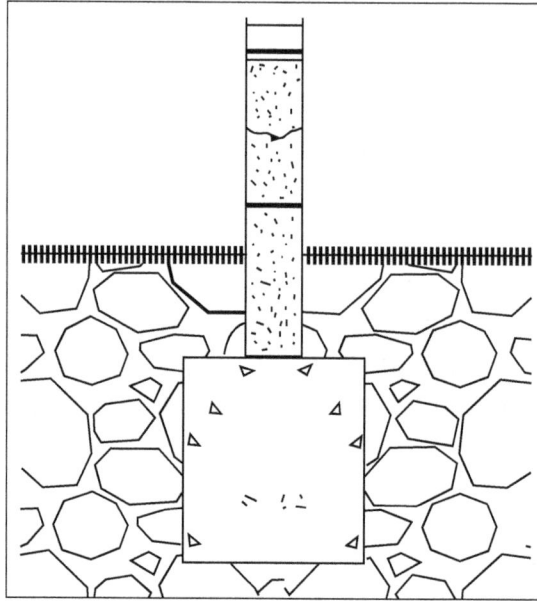

Foundation details.

When the investigation report is not available, the depth of foundation should be selected such that it is not affected by swelling and shrinking of soil due to seasonal changes. A depth of foundation should also consider the depth of water table to prevent and scour below the ground.

For foundation near existing foundation, it should be ensured that pressure bulbs of foundations do not coincide if the depth of new foundation has to be taken below the depth of existing foundation. The foundation should not be constructed at shallow depth considering the frost action in cold countries.

Rankine's formula provides the guidance on minimum depth of foundation based on bearing capacity of soil is,

$$h = \frac{p}{N}\left(\frac{1-\sin\phi}{1\ \sin}\right)^2$$

Where,

h = Minimum depth of foundation.

p = Gross bearing capacity.

γ = Density of soil.

φ = Angle of repose or internal friction of soil.

The guidance on minimum foundation depth, assuming that the foundations are not affected by factors such as water table, frost action, types and properties of soil etc. This formula does not consider the loads from the structure on the foundation.

In the Rankine's formula, it can be use that foundation depth depends on the bearing capacity of soil, so if the bearing capacity of soil increases, the depth of foundation also increases.

Foundation Depth Calculator

Gross bearing capacity of soil	p	300	kN/m²
Density of soil	γ	18	kN/m³
Angle of repose	φ	30	degree
Minimum foundation Depth	h	0	m

2.3   Codal Provisions

Location and Depth of Foundation as Per IS: 1904-1986

The following considerations are necessary for deciding the location and depth of foundation as per IS: 1904-1986, minimum depth of foundation shall be 0.50 m:

- The frost heave.

- Excessive volume change due to moisture variation. It is usually exists within 1.5 to 3.5 m depth of soil from the top surface.

- Topsoil or organic material.

- Peat and Muck.

- Unconsolidated material such as waste dump.

The following steps to be taken for design in such conditions:

- Determine foundation type.

- Estimate cost of foundation for normal and different scour conditions.

- Determine the scour versus risk and revise the design accordingly.

- Estimate probable depth of scour, effects, etc.

IS:1904-1986 recommendations for foundations adjacent to slopes and existing structures When the ground surface slopes downward adjacent to footing, the sloping surface should not cut the line of distribution of the load (2H:1V):

- In granular soils, the line joining the lower adjacent edges of upper and lower footings shall not have a slope steeper than 2H: 1V.

- In clayey soil, the line joining the lower adjacent edge of the upper footing and the upper adjacent edge of the lower footing should not be steeper than 2H: 1V.

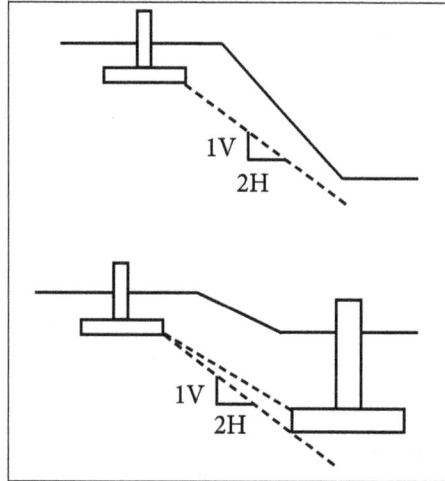

Other Recommendations for Footing Adjacent to Existing Structures

- The minimum horizontal distance between the foundations shall not be less than the width of larger footing to avoid damage to existing structure.

- If the distance is limited, the principal of 2H: 1V distribution should be used so as to minimize the influence to old structure.

- The proper care is needed during excavation phase of foundation construction beyond normally depending on the 2H: 1V criteria for old foundations. Excavation may cause settlement to old foundation due to lateral bulging in the excavation or shear failure due to reduction in overburden stress in the surrounding of old foundation.

Footings on Surface Rock or Sloping Rock Faces

- For the locations with shallow rock beds, the foundation can be laid on the rock surface after chipping the top surface.

- If the rock bed has some slope, it may be advisable to provide dowel bars of minimum 16mm diameter and 225 mm embedment into the rock at 1 m spacing.

A raised water table may cause damage to the foundation by:

- Floating the structure.

- Reducing the effective stress beneath the foundation. Water logging around the building may also cause wet basements. In such a cases, proper drainage system around the foundation may be required so that water does not accumulate.

2.4 Bearing Capacity of Shallow Foundation on Homogeneous Deposits

The bearing capacity is the power of foundation soil to hold the forces from the super-structure without undergoing shear failure or excessive settlement.

A foundation of soil is that portion of ground which is subjected to additional stresses when foundation and superstructure are constructed on the ground. The following are a few important terminologies related to bearing capacity of soil.

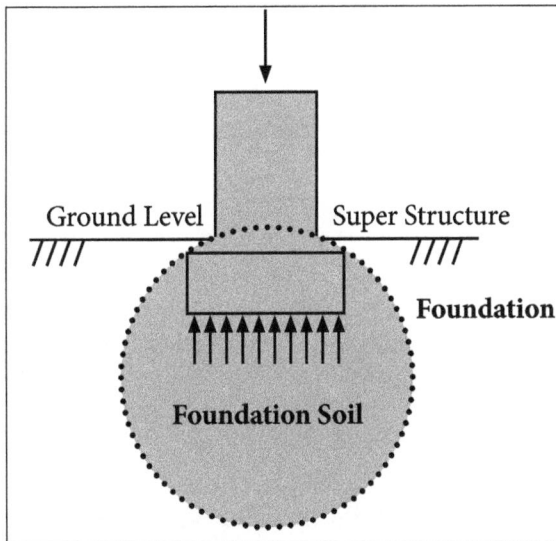

Pressure acting on the foundation soil.

Typical Values of Soil Bearing Capacity

Soil type	Bearing value (kPa)	Remarks
Dense gravel or dense sand and gravel	> 600	Width of foundation not less than 1 m. Water table at least at the depth equal to the width of foundation, below base of foundation.
Dense dense gravel or medium dense sand and gravel	200-600	-

Loose gravel or loose sand and gravel	< 200	-
Compact sand	> 300	-
Medium dense sand	100 to 300	-
Very stiff boulder clays and hard clays	300 to 600	Susceptible to long term consolidation settlement
Stiff clays	150 - 300	-
Soft clays and silts	< 75	-
Firm clays	75 -150	-
Very soft clays and silts	-	-

Ultimate Bearing Capacity (qf)

It is the maximum pressure that a foundation soil can withstand without undergoing shear failure.

Net Ultimate Bearing Capacity (qn)

It is the maximum extra pressure that a foundation soil can withstand without undergoing shear failure,

$$q_n = q_f - q_o$$

Here, q_o represents the overburden pressure at foundation level and is equal to γD for level ground without surcharge where γ is the unit weight of soil and D is the depth to foundation bottom from Ground Level.

Safe Bearing Capacity (qs)

It is the safe extra load at which foundation soil is subjected, in addition to initial overburden pressure,

$$q_s = \frac{q_n}{F} + q_o$$

Here F represents the factor of safety.

Allowable Bearing Pressure (qa)

It is the maximum pressure the foundation soil is subjected to considering both shear failure and settlement.

Criteria for Determination of Bearing Capacity

The criteria for the determination of bearing capacity of a foundation, which are based on the requirements for stability, are as follows:

- Shear failure or bearing capacity failure as it is sometimes called, shall not occur. This is associated with plastic flow and lateral expulsion of the soil beneath the foundation.

- The anticipated settlements, differential as well as total, of the foundation shall be less than the safe or tolerable values, so that the stability of the structure is not impaired.

These two criteria are independent of each other. The design value of the safe bearing capacity is obviously the smaller of the values obtained from these two criteria. This has already been defined as the allowable bearing pressure.

Bearing capacity is governed by a number of factors.

The following are some of the more important factors:

- The nature of soil and its properties.

- The foundation type and details such as size, shape, depth below the ground and rigidity of the structure.

- The total and differential settlements that the structure can withstand without functional failure.

- The depth of ground water table relative to the foundation level.

The ultimate bearing capacity of a footing in a cohesion less soil can be obtained by,

$$q_{nu} = q\,N_q + 0.5\,YBN_Y$$

A sandy soil of equation will identify the factors that have an influence on the bearing capacity. These are:

Relative density or angle of shearing resistance:

- Width of footing.

- Depth of footing.

- Unit weight of soil.

- Position of groundwater table.

The higher the relative density, the greater the angle of shearing resistance, φ' of a cohesion less soil deposit.

Step by Step procedure for the calculation of Bearing Capacity from Standard Penetration Test Values or N-values:

- Step 1: Perform standard penetration test on the location for which we want to calculate bearing capacity. This is done as per standard procedure given in IS-2131. The standard penetration test must be done at every 75 cm in vertical direction.

- Step 2: Decide the width, depth and length of foundation for initial calculation. This is a trial and error process. In the first attempt, we can never get the exact size of foundation which will satisfy all of needs.

- Step 3: Apply necessary corrections to the standard penetration test values. Calculate the cumulative average value of corrected SPT values from the base level of foundation to a depth equal to 2 times the width of foundation.

- Step 4: Correlate the above cumulative average SPT value with the figure given below to find out the corresponding angle of shearing resistance (φ).

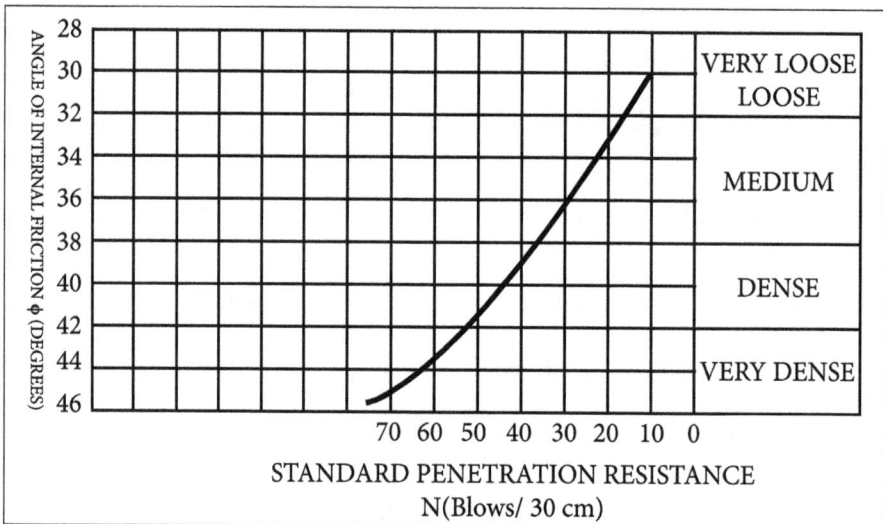

Relation between phi and SPT value (N).

- Step 5: Calculate effective surcharge at the base level of foundation by multiplying the effective unit weight of soil with the depth of the foundation i.e.,

$$q = \gamma * D_f$$

Where,

q = Effective surcharge at the base level of foundation, in kgf/cm².

γ = Unit weight of soil, in kgf/cm³.

D_f = Depth of foundation, in cm.

- Step 6: For the angle of shearing resistance value (φ) as calculated in step-4, find out the corresponding values of bearing capacity factors (i.e. N_q & N_γ) from the table given below. For the intermediate values of 'φ', make linear interpolation.

Bearing Capacity Factors

φ (Degrees)	N_c	N_q	N_γ
0	5.14	1.00	0.00
5	6.49	1.57	0.45
10	8.35	2.47	1.22
15	10.98	3.94	2.65
20	14.83	6.40	5.39
25	20.72	10.66	10.88
30	30.14	18.40	22.40
35	46.12	33.30	48.03
40	75.31	64.20	109.41
45	138.88	134.88	271.76
50	266.89	319.07	762.89

Step 7: Calculate shape factors (i.e. s_q & s_γ) using formula given below:

Shape factors				
Sl.no	Shape of base	s_c	s_q	s_γ
i)	Continuous strip	1.00	1.00	1.00
ii)	Rectangle	1 + 0.2 B/L	1 + 0.2 B/L	1 - 0.4 B/L
iii)	Square	1.3	1.2	0.8
iv)	Circle	1.3	1.2	0.6

Where,

B = Width of foundation, in cm.

L = Length of foundation, in cm.

- Step 8: Calculate depth factors (i.e. d_q & d_γ) using following formula,

$$d_q = d_\gamma = 1 \ \left(\text{for } \phi < 10°\right)$$

$$d_q = d_\gamma = 1 + 0.1\left(D_f / B\right)\left(N_\phi\right)^{1/2} \ \left(\text{for } \phi > 10°\right)$$

N_ϕ is calculated using following formula,

$$N_\phi = \tan^2\left[(\pi/4) + (\phi/2)\right]$$

- Step 9: Calculate inclination factors (i.e. i_q & i_γ) using the formula given below,

$$i_q = \left(1 - \frac{\alpha}{90}\right)^2$$

$$i_\gamma = \left(1 - \frac{\alpha}{\phi}\right)^2$$

Where,

α = Inclination of the load to the vertical in degrees.

ϕ = Angles of shearing resistance in degrees.

- Step 10: Calculate the correction factor for location of water table using the following formula,

$$W' = 0.5 + 0.5 \left[D_w / (D_f + B)\right]$$

Where,

W' = Correction factor for location of water table.

D_w = Depth of water table, in cm.

D_f = Depth of foundation, in cm.

- Step 11: Using the equation given below calculate the net ultimate bearing capacity,

$$d = q\left(N_q - 1\right)S_q\, d_q\, i_q + \frac{1}{2} B\gamma\, N_\gamma s_\gamma\, d_\gamma\, i_\gamma\, W$$

Where,

q_d = Net ultimate bearing capacity of foundation, kgf/cm².

q = Effective surcharge at base level of foundation, in kgf/cm².

N_q & N_γ = Bearing capacity factors.

s_q & s_γ = Shape factors.

d_q & d_γ = Depth factors.

i_q & i_γ = Inclination factors.

W' = Correction factor for location of water table.

B = Width of foundation, in cm.

γ = Bulk unit weight of foundation soil, in kgf/cm^3.

2.5 Terzaghi's Formula and BIS Formula

Terzaghi (1943) was the first to propose a comprehensive theory for evaluating the safe bearing capacity of shallow foundation with rough base.

Assumptions

- Soil is homogeneous and Isotropic.

- The shear strength of soil is represented by Mohr Coulombs Criteria.

- The footing is of strip footing type with rough base. It is essentially a two dimensional plane strain problem.

- Elastic zone has straight boundaries inclined at an angle equal to the horizontal.

- Failure zone is not extended above, beyond the base of the footing. Shear resistance of soil above the base of footing is neglected.

- Method of superposition is valid.

- Passive pressure force has three components (P_{PC} produced by cohesion, P_{Pq} produced by surcharge and $P_p \gamma$ produced by weight of shear zone).

- Effect of water table is neglected.

- Footing carries concentric and vertical loads.

- Footing and ground are horizontal.

- Limit equilibrium is reached simultaneously at all points. Complete shear failure is mobilized at all points at the same time.

- The properties of foundation soil do not change during the shear failure.

Limitations

- The theory is applicable to shallow foundations.

- As the soil compresses, increases which is not considered. Hence fully plastic zone may not develop at the assumed.

- All points need not experience limit equilibrium condition at different loads.

- Method of superstition is not acceptable in plastic conditions as the ground is near failure zone.

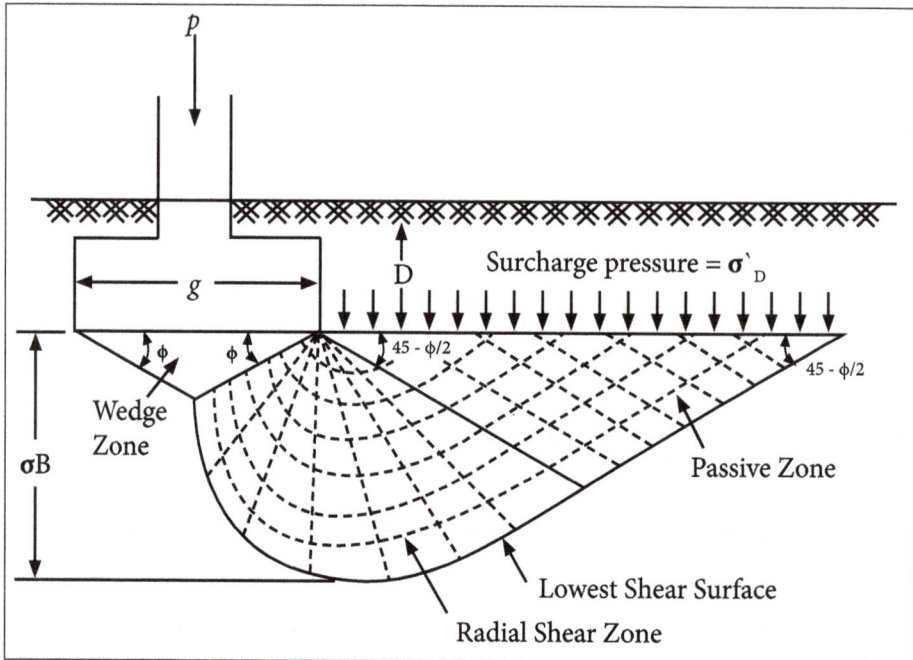

Terzaghi's concept of Footing with five distinct failure zones in foundation soil.

Concept

A strip footing of width B gradually compresses the foundation soil underneath due to the vertical load from superstructure. Let us q_f be the final load at which the foundation soil experiences failure due to the mobilization of plastic equilibrium.

The foundation soil fails along the composite failure surface and the region is divided in to five zones, Zone 1 which is elastic, the numbers of Zone 2 which are the zones of radial shear and two zones of Zone 3 which are the zones of linear shear.

Considering the horizontal force equilibrium and incorporating empirical relation, the equation for ultimate bearing capacity is obtained as follows:

Ultimate bearing capacity,

$$q_f = cN_c + \gamma DN_q + 0.5\,\gamma BN_\gamma$$

If the ground is subjected to additional surcharge load q then,

$$q_f = cN_c + (\gamma D + q)N_q + 0.5\,\gamma BN_\gamma$$

Net ultimate bearing capacity,

$$q_n = cN_c + \gamma D N_q + 0.5\,\gamma B N_\gamma - \gamma D$$

$$q_n = cN_c + \gamma D\left(N_q - 1\right) + 0.5\,\gamma B N_\gamma$$

Safe bearing capacity,

$$q_s = \left[cN_c + \gamma D\left(N_q - 1\right) + 0.5\,\gamma B N_\gamma\right]\frac{1}{F} + \gamma D$$

Here,

> F = Factor of safety (usually 3).
>
> c = Cohesion.
>
> γ = Unit weight of soil.
>
> D = Depth of foundation.
>
> q = Surcharge at the ground level.
>
> B = Width of foundation.
>
> N_c, N_q, N_γ = Bearing Capacity factors.

Bearing Capacity Factors for Different ϕ

φ	N_c	N_q	N_g	N'_c	N'_q	N'_g
0	5.7	1.0	0.0	5.7	1.0	0.0
5	7.3	1.6	0.5	6.7	1.4	0.2
10	9.6	2.7	1.2	8.0	1.9	0.5
15	12.9	4.4	2.5	9.7	2.7	0.9
20	17.7	7.4	5.0	11.8	3.9	1.7
25	25.1	12.7	9.7	14.8	5.6	3.2
30	37.2	22.5	19.7	19.0	8.3	5.7
34	52.6	36.5	35.0	23.7	11.7	9.0
35	57.8	41.4	42.4	25.2	12.6	10.1
40	95.7	81.3	100.4	34.9	20.5	18.8
45	172.3	173.3	297.5	51.2	35.1	37.7
48	258.3	287.9	780.1	66.8	50.5	60.4
50	347.6	415.1	1153.2	81.3	65.6	87.1

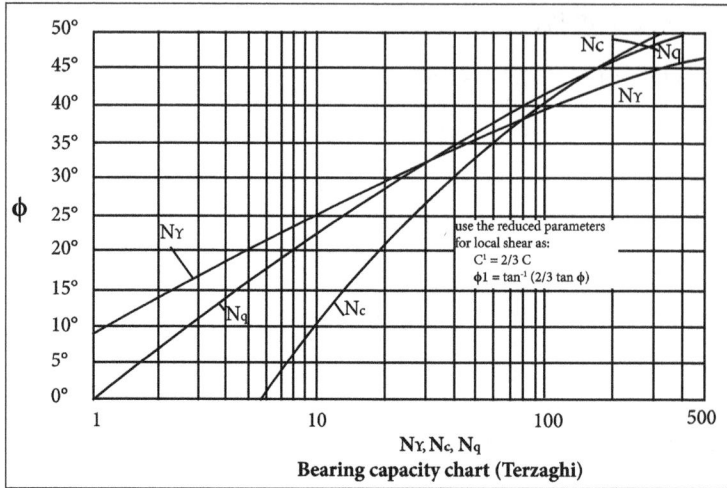

Bearing capacity chart (Terzaghi)

Terzaghi's Bearing Capacity Factors for different ϕ.

IS Methods

IS: 6403-1981 gives the equation for the net ultimate bearing capacity as,

$$q_{nu} = cN_cS_cd_ci_c + q(N_q - 1)s_qd_qi_q + 0.5\gamma \, BN_\gamma s_\gamma d_\gamma i_\gamma W'$$

The factor W' takes into account, the effect of the water table. If the water table is at or below a depth of Df + B, measured from the ground surface, W'=1. If the water table rises to the base of the footing or above, W'=0.5. If the water table lies in between then the value is obtained by linear interpolation. The shape factors given by Hansen and inclination factors as given by Vesica are used. The depth factors are given below,

$$d_c = 1 + 0.2(D_f / B)\tan(45 + \phi'/2)$$
$$d_q = d_\gamma = 1 \text{ for } \phi' < 10°$$
$$D_\gamma = d_q = 1 + 0.1(D_f / B)\tan(45° + \phi'/2) \text{ for } \phi' > 10°$$

For cohesive soils,

$$q_{nu} = cN_cs_cd_ci_c$$

Where N_c = 5.14 and s_c, d_c and i_c are respectively the shape, depth and inclination factors.

Brinch Hansen's Bearing Capacity Equations

Bearing capacity depends on several factors and Terzaghi' s bearing capacity equation does not take in to consideration all the factors. Brinch Hansen and more other researchers have provided a comprehensive equation for the determination bearing capacity known as Generalized Bearing Capacity equation considering the almost all the factors mentioned above. The equation for ultimate bearing capacity is as follows from the comprehensive theory,

$$q_f = cN_c s_c d_c i_c + qN_q s_q d_q i_q + 0.5\,\gamma BN_\gamma s_\gamma d_\gamma i_\gamma$$

Here, the bearing capacity factors are given by the subsequent expressions that depend on φ,

$$N_c = \left(N_q = 1\right) \cot\phi$$

$$N_q = \left(e^{\pi\tan\phi}\right) \tan^2\left(45+(\phi/2)\right)$$

$$N_\gamma = 1.5\left(N_q - 1\right)\tan\phi$$

Equations are available for shape factors $\left(s_c, s_q,\ s_\gamma\right)$, depth factors $\left(d_c, d_q,\ d_\gamma\right)$ and load inclination factors $\left(i_c, i_q,\ i_\gamma\right)$. The effects of those factors is to reduce the bearing capacity.

Bearing Capacity of Footing Subjected to Eccentric Loading

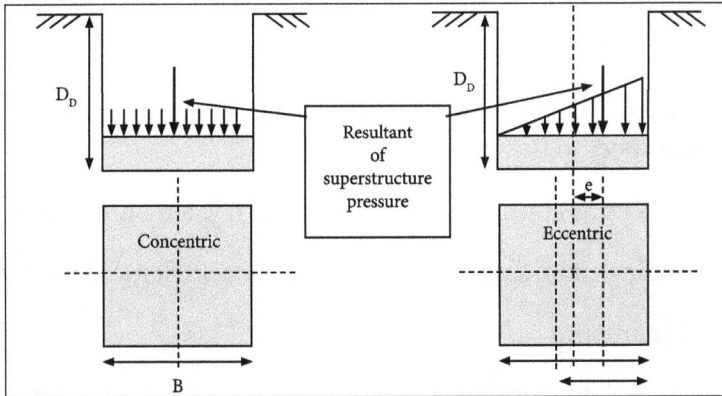

Effect of eccentric footing on bearing capacity.

The bearing capacity equation is developed with the idealization that the load on the foundation is concentric. Although the forces on the foundation could be eccentric or foundation might be subjected to additional moment. In such as situations, the width of foundation B shall be considered as follows,

$$B^1 = B - 2e$$

If the loads are eccentric in each the directions then,

$$B^1 = B - 2e_B \ \& \ L^1 = L - 2e_L$$

Further, area of foundation to be considered for safe load carried by foundation is not the actual area, but the effective area as follows,

$$A^1 = B^1 X L^1$$

In the calculation of bearing capacity, width to be considered is B^1 wherever $B^1 < L^1$. Hence the effect of provision of eccentric footing is to reduce the bearing capacity and load carrying capacity of footing.

2.6   Factors Affecting Bearing Capacity

1. Pre-Earthquake Conditions

Observations of foundation behavior during the 1985 Michoacan- Guerero earthquake clearly evidenced that the initial static pressure and load eccentricity have pronounced effects on the seismic behavior of foundations. Foundations with low static safety factors (high contact pressures) or with significant load eccentricities behaved poorly whereas well-designed foundations appear not to have suffered significant damages.

2. Seismic Toads from the Superstructure

In the general case, the dynamic forces have six components:

- A vertical force which, in most cases, can be neglected since its magnitude is small with respect to the static one.

- Two shear forces (T) inducing an inclination of the resultant force.

- Two overturning moments (M) inducing an eccentricity of the resultant force.

- A torsional moment.

For strip foundations, these forces reduce to the vertical one, the shear force and the overturning moment. It will become apparent later that generally, only the maximum value of these forces are needed but, in some instances, when permanent displacements are to be computed, their variations in time are also required.

3. Soil Strength

Even for static conditions, the correct evaluation of the soil strength is a challenge for the geotechnical engineers. It depends on many factors such as the soil fabric, stress and strain history, stress path, drainage conditions, anisotropy, etc. Additional factors must be considered for seismic conditions, depending on the soil type which will he broadly characterized as cohesion less soils or cohesive soils:

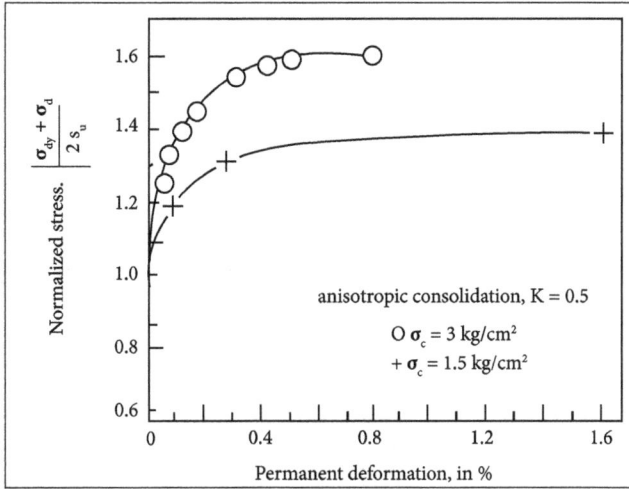

Dynamic stress plastic deformation relationship.

The rate of loading may significantly affect the value of the strength measured in conventional laboratory test usually carried out at slower rates than those anticipated in the field during an earthquake. The strength of cohesion less soils is not affected by the rate of loading but plastic cohesive soils may exhibit untrained strengths 30% to 60% higher than the conventional strength. Figure presents a typical example of such an increase measured or, a clay sample from Mexico city - degradation under cyclic loading.

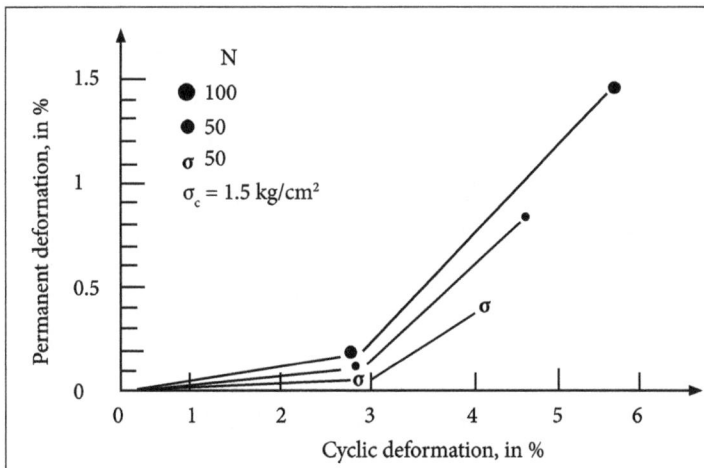

Effect of Cyclic deformation on Permanent deformation.

The repetition of alternate cycles of loading may cause a cyclic degradation of the material and a subsequent decrease in its strength. However, not all materials are subjected to degradation: cohesion less soils are insensitive to it and cohesive soils will only experience degradation when they are strained beyond a given strain threshold which is material dependent. Figure shows, for instance, that permanent cyclic deformations start to accumulate for Mexico city clays only when the cyclic strain is higher than approximately 3%.

Pore pressure build-up and drainage conditions. Saturated cohesion less soils usually experience an increase in their pore water pressure due to cyclic loading under unchained conditions, which may lead to a liquefaction condition unless the drainage conditions allow for a rapid dissipation of these pressures. Whether the soil is initially in a dense or in a loose state will lead to completely different situations: a loose sand loses its whole strength and gives rise to large deformations arid catastrophic failures; dense sands, owing to their dilatant behavior, can mobilize very large undrained strengths, provided drainage is prevented and cannot develop dramatic failures.

4. Inertia Forces in the Soil Mass

During an earthquake, the passage of seismic waves gives rise to inertia forces within the soil volume, which are equilibrated by dynamic stress. These stresses mobilize a fraction of the available soil strength and consequently, the strength available to balance the inertia forces arising from the superstructure is not necessarily the full soil strength. As an extreme case, when the ground acceleration at a given depth is too large, the induced seismic stresses may cause failure this phenomenon called fluidization, has been noticed in numerous theoretical studies.

However, it must be noticed that the inertia forces within the soil mass and the forces acting on the foundation, arising from the inertia forces of the superstructure, both vary in time and there is no obvious reason why they should be in phase and maximum, with the same direction, at the same instant. Moreover, inertia forces within the soil profile decrease with depth due to the attenuation of accelerations, which is not reflected in the aforementioned studies. To properly account for the influence of the soil inertia forces, they must be treated as an independent parameter.

To consider the same seismic coefficient k_H for the soil inertia force $\left(F_x = \rho\, g\, k_H\right)$ and for the structural inertia forces $(T = K_H\, mg)$ leads to erroneous conclusions with respect to the influence of the soil inertia forces; the major reduction in the foundation bearing capacity outcomes from the load inclination and eccentricity on the foundation and not from the soil inertia forces.

Factors Influencing the Choice of Foundation

Principal Factors

- Function of structure.
- Load on foundation.
- Subsurface condition.
- Cost.

Other Factors

- Need of client or owner.

- Type of super structure.

- Environmental considerations.

- Margin of safety.

- Risk Level.

- Service life.

2.7   Bearing Capacity from In-Situ Tests

Plate Load Test (IS 1888:1982)

It is a field test to determine the ultimate bearing capacity of soil and the probable settlement under a given loading. The test essentially consists of loading a rigid plate at the foundation level and determining the settlements corresponding to each load increment. The ultimate bearing capacity is then taken as the load at which the plate starts sinking at a rapid rate. The method assumes that down to the depth of influence of stresses and the soil strata is reasonably uniform.

Bearing Plate

Circular or square bearing plates of mild steel, not less than 25 mm in thickness and varying in size from 300 to 750 mm with chequered or grooved bottom, provided with handles for convenient setting and center marked. As an alternative, cast in-situ or pre-cast concrete blocks may be used with depths not less than two-third the width.

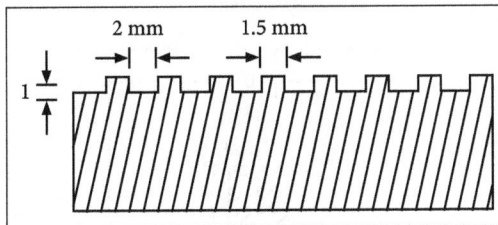

Details of chequers or grooves in test plate.

Size and Shape of Plate

Except in case of road problems and circular footing. Square plates should be adopted. For the clayey and silty soils and for loose to medium dense sandy soils with N < 15, a 450 mm square plate or concrete blocks shall be used.

In the case of dense sandy or gravelly soils (15 < N < 30) three plates size of 300 mm to 750 mm shall be used depending upon practical considerations of reaction loading and maximum grain size. The side of the plate shall be at least four times maximum size of the soil particles present at the test location.

Test Pit

The pits, usually at the foundation level, having in general normally of width equal to five times the test plate or block, shall have a carefully leveled and cleaned bottom at the foundation level and protected against disturbance or changes in natural formation.

Loading Arrangement

The loading to the test plate may be applied with the help of a hydraulic jack.

The reaction of the hydraulic jack may be borne by either of the following two methods:

- Gravity loading platform method.

- Reaction truss method.

1. Gravity Loading Method

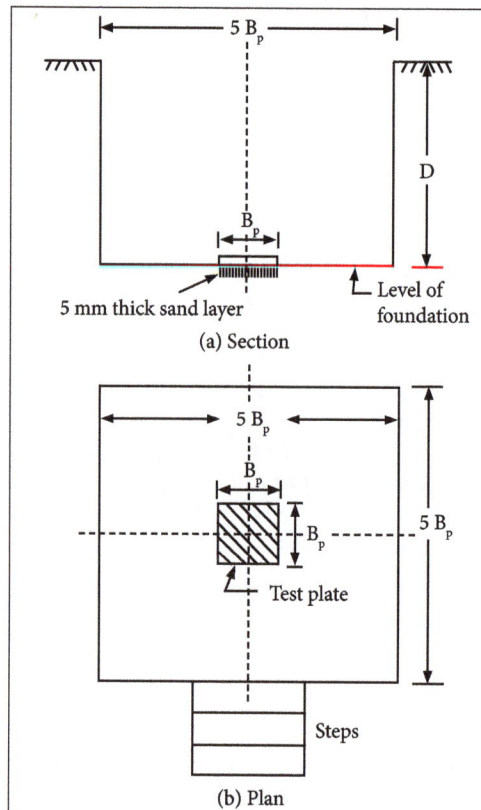

(a) Section

(b) Plan

Gravity loading method.

In the case of gravity loading method, a platform is constructed over a vertical column resting on the test plate. The loading is done with the help of sand bags, stones or concrete blocks. The general arrangement of the test set-up for this method is shown in figure below.

When load is applied to the plate, it sinks or settles. The settlement of the plate is measured with the sensitive dial gauges. For the square plate, the used in two dial gauges. The dial gauges are mounted on independently supported datum bar. As the plate settles, the ram of the dial gauge moves down and settlement is recorded. The load is indicated on the load-gauge of the hydraulic jack.

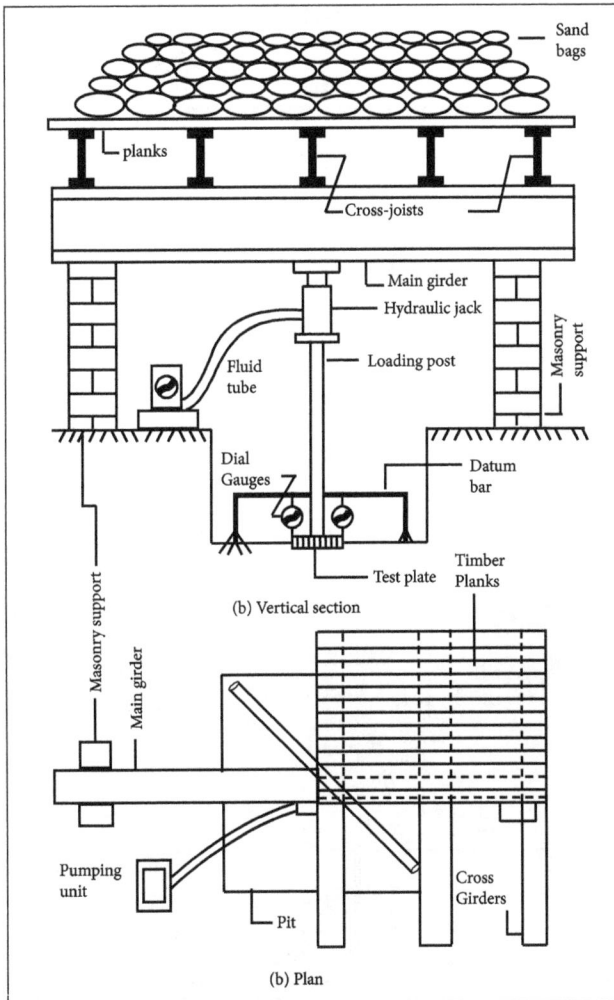

Plate load test: Reaction by gravity loading.

2. Reaction Truss Method

Figure shows the arrangement when the reaction of the jack is borne by are action truss. The truss is held to the ground through soil anchors. These anchors are firmly driven in the soil with the help of hammers. The reaction truss is commonly made of mild steel sections. Guy ropes are used for the lateral stability of the truss.

Indian Standard Code (IS: 1888-1982) recommends that the loading of the plate should invariably be borne either by gravity loading platform or by the reaction truss.

The use of the reaction truss is more popular now-a-days since this is simple, quick and less clumsy. No support of loading platform should be located within a distance for 3.5 times the size of the test plate from its center.

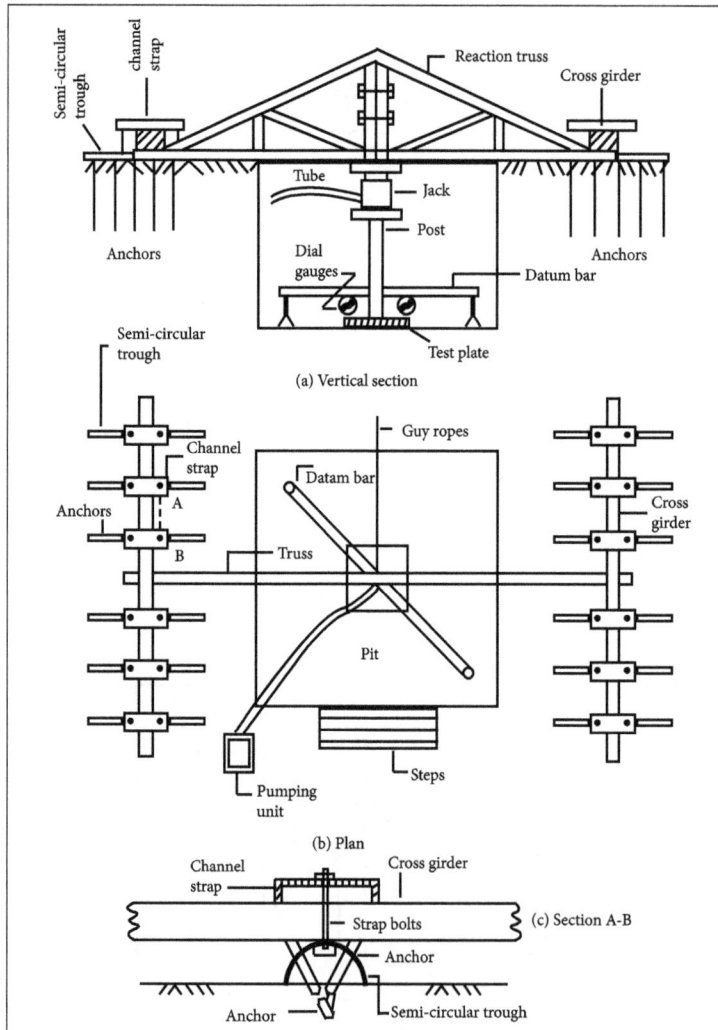

Plate load test: Reaction by truss.

Setting of the Plate

The test plate shall be placed over a fine sand layer of maximum thickness 5 mm. The center of the plate coincides with the center of the reaction girder/beam, with the help of a plumb and bob. Horizontally leveled by the spirit level to avoid eccentric loading.

The hydraulic jack should be centrally placed over the plate with the loading column in between the jack and the reaction beam so as to transfer the load to the plate.

A ball and socket arrangement shall be inserted to keep the direction of the load vertical throughout the test. A minimum seating pressure of $70g/cm^2$ (0.7 t/m^2) shall be applied and removed before starting the load test.

Load Increments

Apply the load to soil in cumulative equal increments up to 1 kg/cm^2 (10 t/m^2) or one fifth of the estimated ultimate bearing capacity, whichever is less. The load is applied without any impact, fluctuation or eccentricity.

In case of hydraulic jack, load is measured over the pressure gauge, attached to the pumping unit kept over the pit, away from the testing plate, through extending pressure pipes.

Settlement and Observations

The settlement should be observed for both increment of load after an interval of 1, 2.25, 4, 6.25, 9, 16 and 25 minutes and thereafter at hourly intervals to the nearest 0.02 mm.

The case of clay soils, the time-settlement curve shall be plotted at both load stage and load shall be increased to next stage either when the curve indicates that the settlement has exceeded 70 to 80 percent of probable ultimate settlement at that stage or at the end of 24 hour period.

For the soils other than clay soils, both load increment shall be kept for not less than one hour or up to a time when the rate of settlement gets appreciably reduced to a value of 0.02 mm/min.

The next increment of load shall then be applied and the observations repeated. The test shall be continued till a settlement of 25 mm under the normal circumstances or 50 mm in special cases such a is dense gravel, gravel and sand mixture is obtained or till failure occurs, whichever is earlier.

When settlement does not reach 25 mm, the test should be continued to at least two times the estimated design pressure. If needed, rebound observations may be taken while releasing the load.

Load Settlement Curve and Ultimate Bearing Capacity

A load settlement curve is plotted out to arithmetic scale. From this load is settlement curve, is zero correction which is given by the intersection of the early straight line or nearly straight line of the curve with zero load line shall be determined and subtracted from the settlement readings to allow for the perfect seating of the bearing plate and other causes. For four typical curves are shown in figure below.

Curve A is typical for loose to medium cohesion less soil: It is a straight line in the earlier stages but flattens out at later stage and there is no clear point of failure.

Curve B is for cohesive soil, it may not be quite straight in the early part and leans towards settlement axis as the settlement increases.

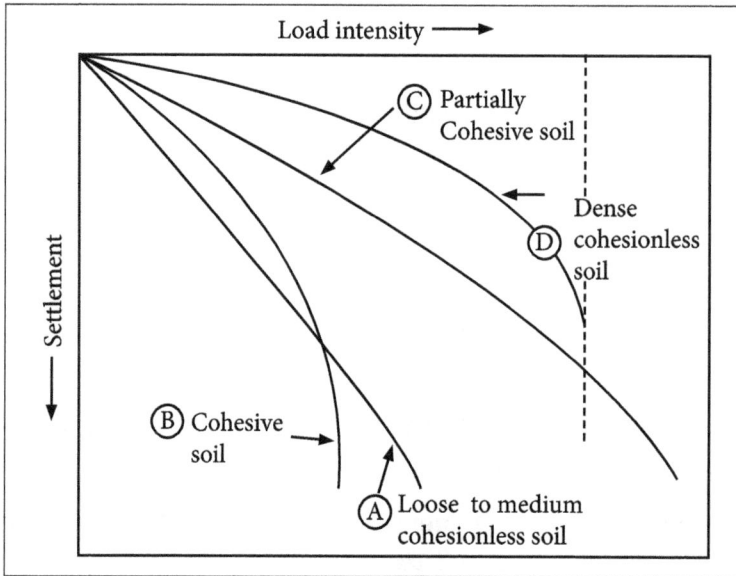

Load settlement curve.

For partially cohesive soils, curve C possessing the characteristics of each curves A and B is obtained, Curve D is purely for dense cohesion less soils.

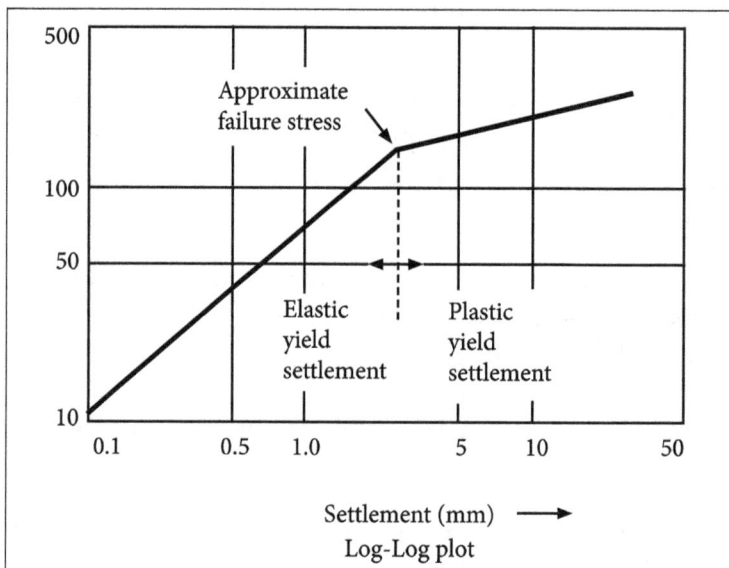

Load settlement curve on log-log plot.

For curves B and D, no difficulty will be experienced in arising at the ultimate bearing capacity as the failure is well defined.

In case of curves A and C, where yield point is not well defined, settlements are plotted as abscissa against corresponding load intensities as ordinate, each to logarithmic scale.

This plotting will give two straight lines, the intersection of which will be considered as yield value of soil.

IS: 1888: 1982 has not specified any factor of safety. In order to determine the safe bearing capacity, it would be normally sufficient to use a factor of safety of 2 or 2.5 on the ultimate bearing capacity.

Limitations of Plate Load Test

The plate loading test has the following:

- The test results reflect only the character of the soil located within a depth less than twice the width of the bearing plate. Since the foundations are commonly larger, the settlement and resistance against shear failure will depend on the properties of a much thicker stratum.

- It is essentially a short duration test and although the test does not give the ultimate settlement, particularly in the case of cohesive soil.

- Another limitation is the effect of the size of foundation. For clayey soils the ultimate pressure for a large foundation is the same as that for the test plate. But in dense sandy soils, the bearing capacity increases, with the size the foundation and the test on smaller size bearing plates tend to given conservative values.

Effect of the Size of Plate on Bearing Capacity

As stated in limitation 3 above, the bearing capacity of sands and gravels increases with the size of the footing. The relationship can be expressed as under,

$$q_F = M + N \frac{B_F}{B_P}$$

The above equation can be solved graphically by using more than one size plates. By extrapolating the plate load test data, one may use the following equation for all practical purposes,

$$q_F = q_P \cdot \frac{B_F}{B_P}$$

Where,

q_F = Bearing capacity of the actual footing.

B_F = Width of actual footing.

q_p = Bearing capacity, obtained from plate load test.

B_p = Width of plate.

However, for clays, the bearing capacity is almost independent of the footing size or the plate size,

$q_f \approx q_P$

For a c–φ soil suggested the following expression,

$Q = A . q + P . s$

Where,

Q = Total load on bearing area.

P = Perimeter of footing.

s = Perimeter shear.

A = Contact area of footing or plate.

q = Bearing pressure beneath area A.

The above equation can be used for the design of a footing to carry a given load, by conducting plate load tests on two different size plates and determining loads Q_1 and Q_2 corresponding to a given fixed settlement.

The test data can then be used in equation to determine q and s. When once q and s are known, equation can be reused to get foundation size to carry a given load Q.

Effect of Size of Plate on Settlements

The settlement of a footing varies with its size. Terzaghi and Peck have suggested the following relationship between the settlement of plate (p_p) and settlement of actual footing (P_F) for granular soils,

$$\rho_P = \rho_F \left[\frac{B_P(B+0.3)}{B(B_P+0.3)} \right]^2$$

If $\Delta \rho$ is the permissible settlement of the foundation, the maximum settlement of the largest footing should be restricted to $\frac{4}{3} \Delta\rho$.

The corresponding settlement of the test plate (ρ_P) on sand, soil is given by,

$$\rho_p = \frac{4}{3}\Delta\rho \left[\frac{B_P(B+0.3)}{B(B_P+0.3)}\right]^2$$

The net loading intensity q_ρ corresponding to ρ_P, determined from the load settlement curve from the plate is then the bearing pressure for a specified settlement for the plate and $\frac{4}{3}\Delta\rho$ for the large footing. For clayey soils, the following relation can be used,

$$\rho_P = \rho_F \cdot \frac{B_P}{B}$$

Bond has proposed the following equation for settlements,

$$\frac{\rho_F / B}{\rho_P / B_P} = \left(\frac{B}{B_P}\right)^n$$

or,

$$\frac{\rho_F}{\rho_P} = \left(\frac{B}{B_P}\right)^{n+1}$$

Where, n = Coefficient depending on the types of soil.

The value of index n can be determined by carrying out two or more plate load tests on different size plates. In absence of test data, the following values of n can be adopted:

- Clay: 0.03 to 0.05.

- Sandy clay: 0.08 to 0.10.

- Medium to dense sand: 0.25 to 0.35.

- Loose sand: 0.20 to 0.25.

- Dense sand: 0.4 to 0.5.

Effect of Ground Water Table on Bearing Capacity

The ground water reduces the ultimate bearing capacity of all soils under the water table. The reason for this is the apparent reduction in soil density owing to the effect of buoyancy. Whether the presence of ground water should be considered in computing the ultimate bearing capacity of any particular footing depends upon the elevation of the water table with respect to the footing.

Such as adverse condition can happens whenever the water table is above an arbitrarily chosen limit line located a distance beneath the footing equal to the width of the footing. This condition and the modifications required by it are indicated in figure below.

It should be noted that ground water could fluctuate seasonally or might be susceptible to short term variation owing to flooding or drought. It is the responsibility of the designer to take those matters into consideration.

Once making calculations, remember, the submerged weight of the soil is,

$$\gamma - 62.4$$

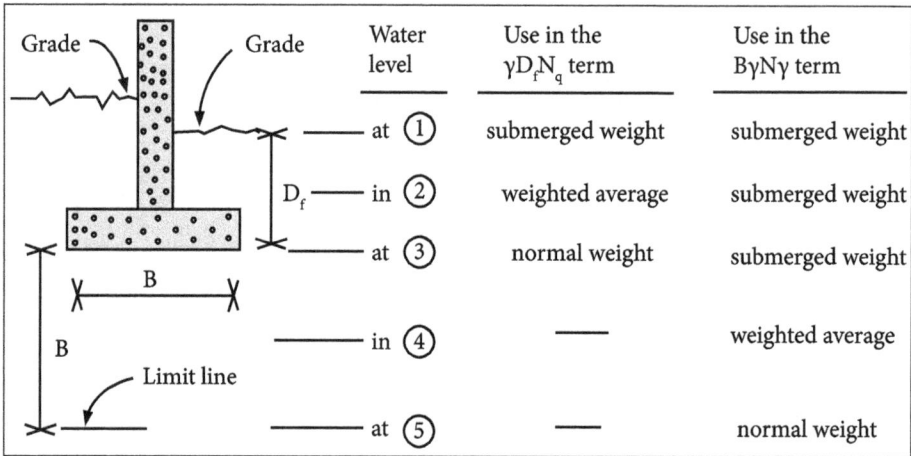

Water level	Use in the $\gamma D_f N_q$ term	Use in the $B\gamma N\gamma$ term
at ①	submerged weight	submerged weight
in ②	weighted average	submerged weight
at ③	normal weight	submerged weight
in ④	—	weighted average
at ⑤	—	normal weight

The reduction in unit weight of a soil, due to pound water.

2.8 Allowable Bearing Pressure

IS: 2950-1965 gives the code of practice for design and construction of raft foundations. The maximum differential settlement should not exceed 40 mm in foundation on clayey soils and 25 mm in foundations on sandy soils. The maximum settlement should generally be limited to the following values:

- Raft foundation on clay: 65 to 100 mm.

- Raft foundation on sand: 40 to 65 mm.

The behaviour of a raft foundation being complicated, a number of simplifying assumptions have to be made in the design. There are two approaches for design - conventional method and the elastic method or the soil line method.

1. Conventional Method

The conventional method is based on the following two basic assumptions:

- The foundation is infinitely rigid and, therefore, the actual deflection of the raft does not influence the pressure distribution below the raft.

- The soil pressure is assumed to be planer such that the centroid of the soil pressure coincides with the line of action of the resultant force of all the loads acting on the foundation.

In the conventional method, the allowable bearing pressure is found using the following formulae,

$$q_1 = 21.4 \, N^2 \, B \, R_{W_1} + 64\left(100 + N^2\right) DR_{w2} \quad ...(1)$$

$$q_2 = 1950 \, (N-3) R_{w2} \quad ...(2)$$

Where,

q_1 and q_2 = Allowable soil pressure under raft foundation in kg/m².

(Using a factor of safety of three),

R_{W_1} and R_{W_2} = Reduction factors on account of sub-soil water.

The smaller of the two values of q_1 and q_2 should be used for design. The effect of overburden pressure on the N value should also be considered. In the case of saturated silts, the equivalent penetration resistance N_e, for the values of N greater than 15 should be taken for design.

The pressure distribution (q) under the raft should be determined by the following formula,

$$q = \frac{Q}{A} \pm Q \frac{e'_y}{I'_x} y \pm Q \frac{e'_x}{I'_y} x$$

Where,

Q = Total vertical load on the raft.

x, y = Co-ordinates of any given point on the raft with respect of the x and y axes passing through the centroid of the area of the raft.

I_x', I_y', e_x', e_y' = Moments of inertia and eccentricities about the principal axes through the centroid of the section,

A = Total area of the raft

$I_{x'}$, I_y', e_x' and e'_y Can be calculated from the following equations:

Where, I_x, I_y = Moment of inertia of the area of the raft respectively about the x and y axes through the centroid,

$I_{xy} = \int_{xy} dA$ for the whole area about x and y axes through the centroid.

e_x, e_y = Eccentricities in the x and y directions of the load from the centroid.

If one or more of the values of q are negative as calculated by the above formula, it indicates that the whole area of foundation is not subjected to pressure and only a part of the area is in contact with soil and the above formula will still hold good provided, the appropriate values of I_x, I_y, e_x and e_y are used with respect to the area in contact with the soil instead of the whole area.

Thus, the pressure distribution is known from equation. The raft is analyzed as a whole in each of the two perpendicular directions.

The total shear force acting on any section cutting across the entire raft is equal to the arithmetic sum of all forces and reaction to the left or right of the section. Similarly the total bending moment acting on such section is equal to the sum of all moments to the left or right of this section.

The structural design of the reinforced concrete raft can now be done, considering the foundation as an inverted floor. The actual reinforcement provided should be twice of that worked out theoretically in the case of design by conventional method.

2.9 Seismic Considerations in Bearing Capacity Evaluation

The recommendations of state that in the calculation of the bearing capacity of a shallow foundation we should include the eccentricity arising from the inertia forces of the structure and the load inclination and also the possible effects of the inertia in the soil.

Hence, the seismic analysis was carried out taking into account the following seismic coefficients: kh_1 = 0.1 and 0.2 for the inertia of the structure. kh_2 = 0.2 and 0.4 for the inertia of the soil mass.

The value of kh1 is chosen lower than kh2 because of the fact that the Euro code 8 allows us to reduce the seismic action by a behaviour factor associated with the ductility classification of the structures.

This consideration leads to the conclusion that the kinematic effect and the consequent reduction in bearing capacity due to the soil inertia, cannot be eliminated and in some circumstances it's reduction can be more significant than the reduction due to the structure's inertia. In this study the seismic coefficient kh_3 of the surcharge was equal to kh_1.

The friction angle of soil was chosen in the range of 0° to 40°, while the angle of the slope near the footing was varied in the range of 5° to 35°.

In the seismic analysis the angle of the sloping ground is affected by further limitations, because simple equilibrium considerations, for a cohesion less soil (c' = 0), take to the following,

$$\tan \beta < \frac{(1-k_v)\tan\phi' - k_{h,i}}{1 - k_v + k_{h,i}\,\tan\phi'} \quad (i = 2,3)$$

$$\beta < \hat{o}' - \theta$$

$$\theta = \tan^{-1}\left[\frac{k_{h,i}}{1-k_v}\right] (i = 2, 3)$$

In the following table shows some limit values of β, for k_v = 0 and φ '= 20°, 30° and 40°.

φ	k_{h2}, k_{h3}			
	0.1	0.2	0.3	0.4
20°	14.29°	8.69°	3.30°	-
30°	24.29°	18.69°	13.30°	8.19°
40°	34.29°	28.69°	23.30°	18.19°

Table Limit values of β for a vertical seismic coefficient k_v = 0.

The results of the parametric analysis are shown in the synthetic form. The seismic bearing capacity ratios $N_c^*/N_c, N_q^*/N_q, N^*/N$ are represented as a function of d/B for different slope angles and for various values of the friction angle of the soil.

The threshold distance (dt) at which the sloping ground does not affect anymore the bearing capacity mainly increases with the increase in the angle of friction and secondarily with the increase in the seismic coefficient and with the increase in the slope angle.

The embedment depth of the footing does not play a significant role on the threshold distance; however it may produce a considerable increase in the bearing capacity.

Referring to the N_c^*/N_c ratios, we can observe the values of the normalized threshold distances varying about dt/B = 1, for an undrained analysis (φu = 0°) and dt/B = 5 for φ= 40°.

For the N_q^*/N_q ratios, we determined the values of the normalized threshold distances varying about dt/B = 2, for φ = 20° and about dt/B = 4 for φ = 40°.

Finally for the N*/N ratios, we determined the values of the normalized threshold distances varying about dt/B = 1.5 for φ = 20° and about dt/B = 4 for φ = 40°.

No significant difference in the threshold distance was found when the inertia of the structure or the inertia of the soil mass is considered. Further, the combined effects of soil and structure inertia can be taken into account by using the principle of superposition of the effects.

In this case, at the same way as found by Paolucci & Pecker (1997) and Cascone (2004), the bearing capacity of the soil self-weight under both the seismic loading due to the coefficients kh_1 and kh_2, can be evaluated through the following equation,

$$q_{lim} = \frac{1}{2} B\gamma \, N_{\gamma e} \approx \frac{1}{2} B\gamma \, N_\gamma \, e_{\gamma i} \, e_{\gamma k}$$

Where,

$N_{\gamma e}$ = Bearing capacity factor reduced by both acting the coefficients k_{h1} and k_{h2}.

N = Static bearing capacity factor.

$e_{\gamma i} = N_{\gamma 1}*/N\gamma$ Bearing capacity ratio for structure inertia only

$\left(k_{h1} > 0, \, k_{h2} = k_{h3} = 0\right)$.

$e_{\gamma k} = N_{\gamma 2}*/N_\gamma$ Bearing capacity ratio for soil mass inertia only

$\left(k_{h2} > 0, \, k_{h1} = k_{h3} = 0\right)$.

2.10 Determination of Settlement of Foundations on Granular and Clay Deposits

The elastic settlement of a shallow foundation can be estimated by using the theory of elasticity. Referring to figure and using Hooke's law,

$$S_e = \int_0^H \varepsilon_z \, dz = \frac{1}{E_s} \int_0^H \left(\Delta p_z - \mu_s \, \Delta p_x - \mu_s \, \Delta p_y\right) dz$$

Where,

S_e = Elastic settlement

E_s = Modulus of elasticity of soil

μ_s = Poisson's ratio of the soil

H = Thickness of the soil layer

Δp_x, Δp_y, Δp_z = Stress increase due to the net applied foundation load in the x, y and z directions, respectively.

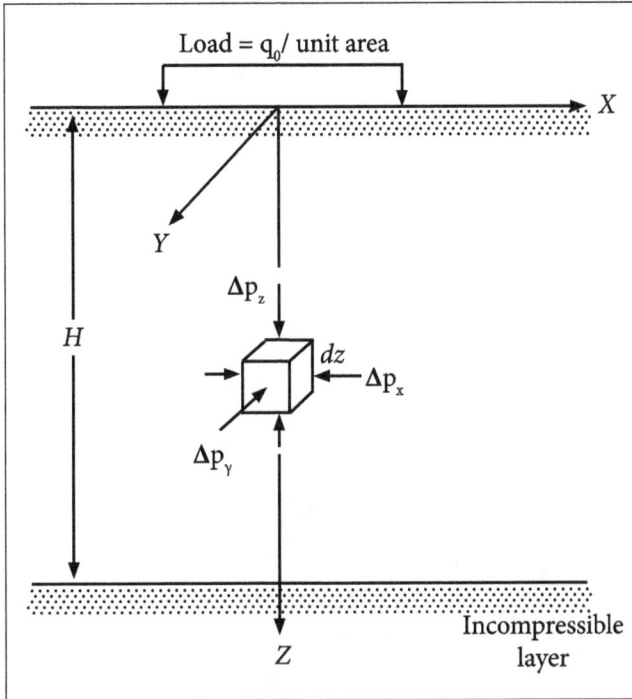

Elastic settlement of shallow foundation.

Theoretically, if the depth of foundation $D_f = 0$, $H = \infty$ and the foundation is perfectly flexible then according to Harr the settlement may be expressed as:

- $S_e = \dfrac{Bq_0}{Es}\left(1-\mu_s^2\right)\dfrac{\alpha}{2}$ (Corner of the flexible foundation).

- $S_e = \dfrac{Bq_0}{Es}\left(1-\mu_s^2\right)\alpha$ (Center of the flexible foundation).

Where,

$$\alpha = \frac{1}{\pi}\left[\ln\left(\frac{\sqrt{1+m_1^2}+m_1}{\sqrt{1+m_1^2}-m_1}\right)+m\ln\left(\frac{\sqrt{1+m_1^2}+1}{\sqrt{1+m_1^2}-1}\right)\right]$$

$m_1 = L/B$

B = Width of foundation.

L = Length of foundation.

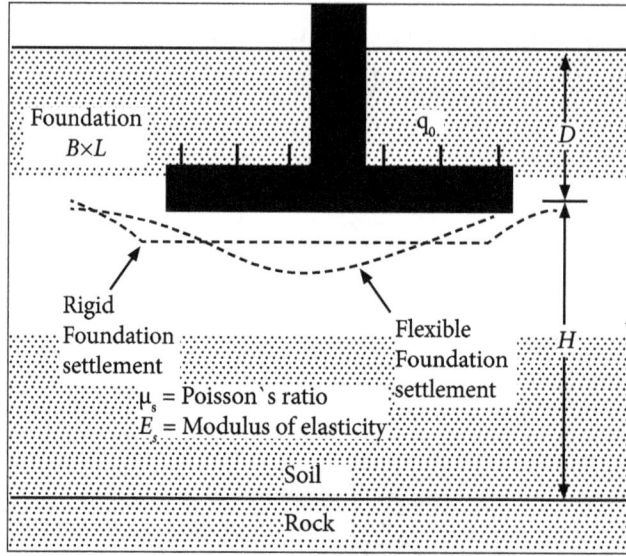

Elastic settlement of flexible and rigid foundations.

The average immediate settlement for a flexible foundation also may be expressed as,

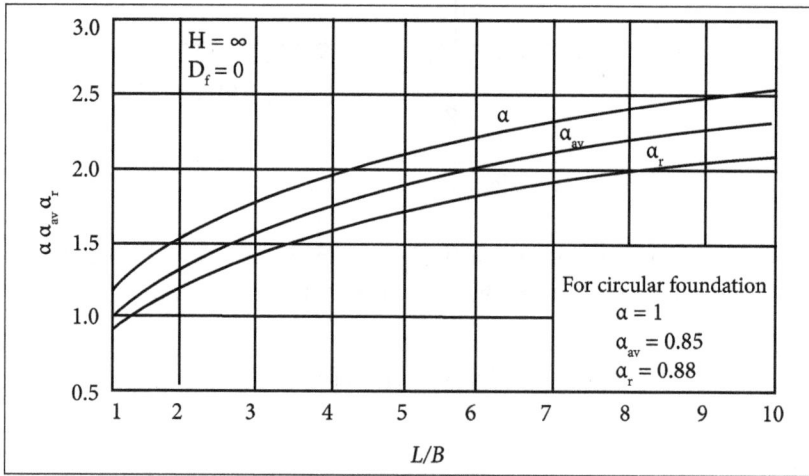

Values of α, α_{av} and α_r –equations.

$$S_e = \frac{Bq_o}{E_s}\left(1-\mu_s^2\right)\alpha_{av} = \text{average for flexible foundation.}$$

Graph also shows the values of α_{av} for various L/B ratios of foundation.

The immediate settlement will be different and may be expressed as,

$$S_e = \frac{Bq_o}{E_s}\left(1-\mu_s^2\right)\alpha_r = \text{(rigid foundation)}$$

The value of αr for various L/B ratios of foundation are shown in above figure.

If $D_f = 0$ and $H < \infty$ due to the presence of a rigid (incompressible) layer,

$$S_e = \frac{Bq_o}{E_s}\left(1-\mu_s^2\right)\frac{\left[\left(1-\mu_s^2\right)F_1 +\left(1-\mu_s -2\mu_s^2\right)F_2\right]}{2} \quad \text{(Corner of flexible foundation)}$$

And,

$$S_e = \frac{Bq_o}{E_s}\left[\left(1-\mu_s^2\right)\left(1-\mu_s^2\right)F_1 +\left(1-\mu_s -2\mu_s^2\right)F_2\right] \quad \text{(Corner of flexible foundation)}$$

The variation of F_1 and F_2 with H/B are given in following graphs respectively:

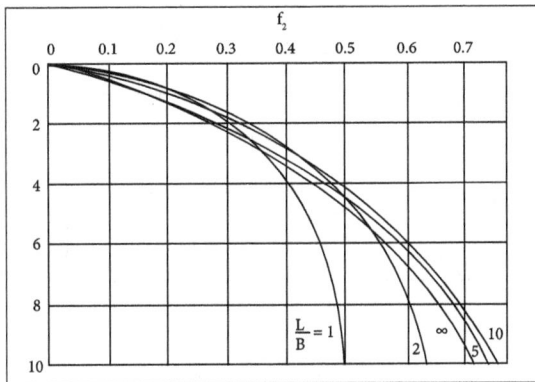

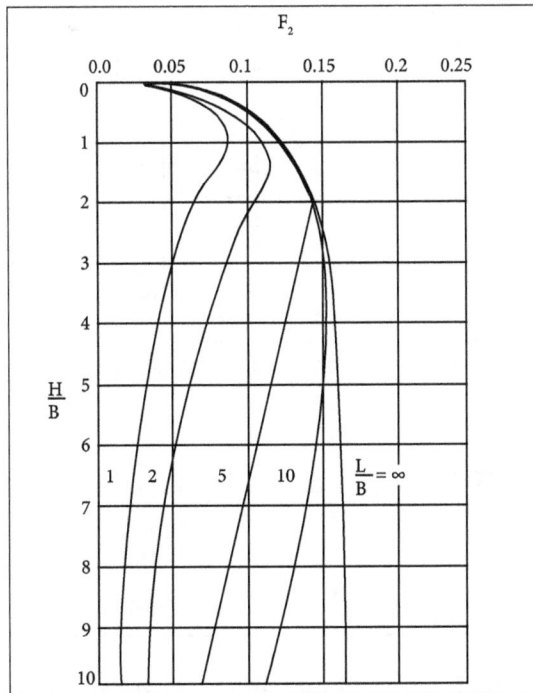

Variation of F_1 and F_2 with H/B.

It is also important to realize that the preceding relationships for S_e assume that the depth of the foundation is equal to zero. For $D_f > 0$, the magnitude of Se will decrease.

Elastic Settlement of Foundation on Saturated Clay

An equation for evaluating the average settlement of flexible foundations on saturated clay soils (Poisson's ratio, $\mu_s = 0.5$). For the notation used in figure, this equation is,

$$S_e = A_1 \, A_2 \, \frac{q_o B}{E_s}$$

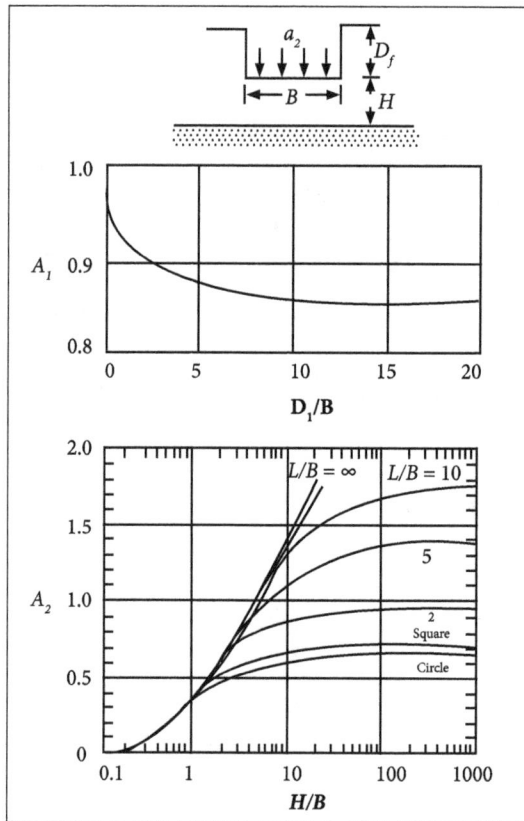

Values of A_1 and A_2 for elastic settlement calculation-equation.

Where,

A_1 is a function of H/B and L/B and A_2 is a function of D_f/B.

2.11 Total and Differential Settlement

Settlement is the vertically downward movement of structure due to the compression of underlying soil because of increased load.

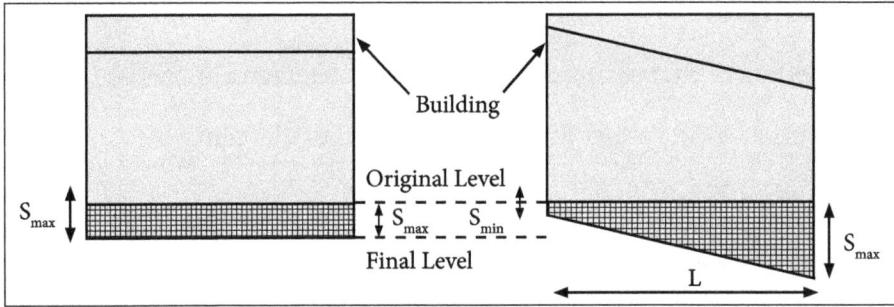

Concepts of uniform and differential settlement.

Differential Settlement

It is the maximum difference between two points in a building element.

Differential Settlement = $S_{max} - S_{min}$

The test results obtained indicate that the differential settlement will be small as long as the average shear stress along the perimeter of the reinforced block is less than the average shear strength of the surrounding soil.

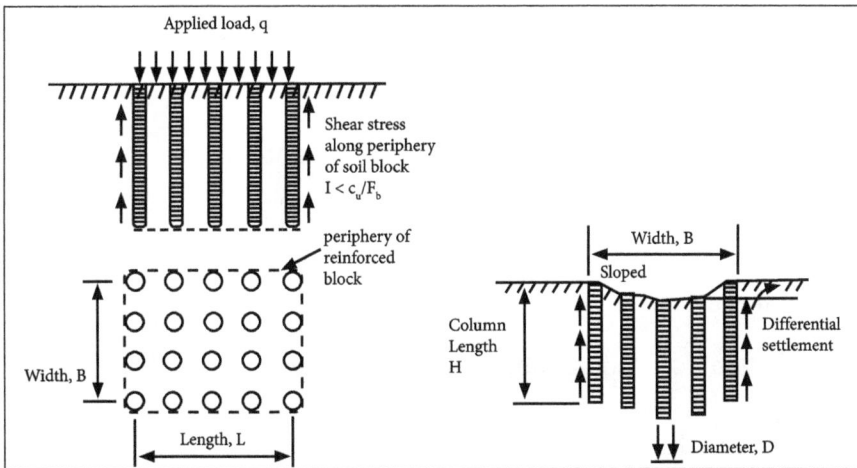

The perimeter shear stress (t) can be calculated from the following equation when the load transferred through the bottom of the stabilized area is neglected,

$$t = W/2(B+L)H < c\,u/f_b$$

Where,

W = Weight of the structure.

f_b = Factor of safety (1.5).

When the load is unevenly distributed, the columns could be concentrated at some parts of the structure with the highest unit load.

Angular Distortion

It is another method of expressing the differential settlement. It can be written as,

Angular Distortion = Differential Settlement/Length of element,

$$= \left(S_{max} - S_{min} \right) / L$$

The components of settlement in soil,

The total settlement of a loaded soil has three components: Immediate settlement, primary consolidation and secondary compression,

S_i - Immediate settlement.

S_c - Consolidation settlement.

S_{sc} - Secondary consolidation settlement.

Total settlement $S_t = S_i + S_c + S_{sc}$.

Immediate Settlement

- Immediate settlement is also known as elastic settlement.
- It is find from elastic theory.
- It happens in all types of soil due to elastic compression.
- It happens immediately after the application of load.
- It depends on the elastic properties of foundation soil, rigidity, size and shape of foundation.

Immediate settlement is calculated by the equation mentioned below,

$$S_1 = \left(\frac{1-\mu^2}{E} \right) q\, BI_\rho$$

Here,

S_I = Immediate settlement.

μ = Poisson's Ratio of foundation soil.

E = Young's modulus of Foundation Soil.

q = Contact pressure at the base of foundation.

B = Width of foundation.

I_ρ = Influence Factor.

Consolidation Settlement

- The consolidation is the process of reduction in volume owing to expulsion of water under an increased load.

- It occurs owing to the process of consolidation.

- It is often confused with Compaction.

- Clay and organic soil are most prone to consolidation settlement.

- It is a time related process happening in saturated soil by draining water from void.

- Consolidation theory is needed to predict both rate and magnitude of settlement.

- Since water flows out in any direction, it is a three dimensional process.

- However, soil is confined laterally. Hence, vertical one dimensional consolidation theory is acceptable.

- Spring analogy explains consolidation settlement.

- Permeability of soil influences consolidation.

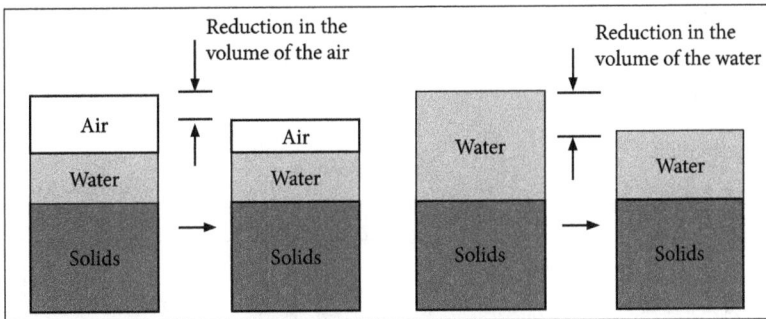

Compaction Consolidation.
Comparison between Compaction and Consolidation.

Method of Estimation

The angle change (α) at the edge of the reinforced block will, at low stress levels, increases approximately linearly with the shear stress (t),

$$\alpha = t / G B$$

G B = Equivalent shear modulus of the soil that depends on the stiffness and the dimensions of the lime columns,

$$G B = B / (B - nD) \times G_{clay}$$

Where,

n = Number of column rows.

D = Column diameter.

G_{clay} = Shear modulus of the unsterilized clay.

Total Settlement

Total foundation settlement is classified into three different components, namely Immediate or elastic settlement, consolidation settlement and creep or secondary settlement as given below,

$$S = S_I + S_C + S_S$$

Here,

S = Total Settlement.

S_I = Immediate / Elastic Settlement.

S_C = Consolidation Settlement.

S_S = Secondary Settlement.

Most structures can tolerate large total settlement if the settlements are distributed evenly. Connecting water and sewer lines begin to break when the total maximum settlement exceeds 150-200 mm.

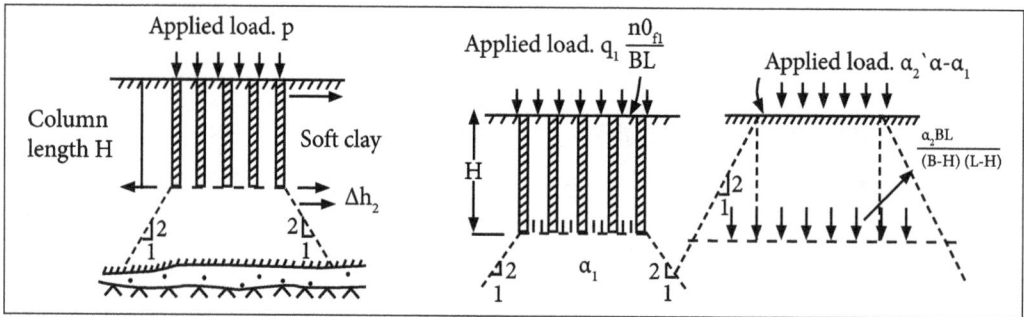

Total settlement.

To calculate the total settlement below the center of the loaded area, the settlement is assumed to be equal to the sum of the compression of the reinforced block (Δh_1) and the compression of the underlying soil (Δh_2).

2.12 Allowable Settlements

The allowable settlement is defined as the acceptable amount of settlement of the structure and it usually includes a factor of safety. This settlement depends on many factors, including the following:

Type of Construction: For example, wood-frame buildings with wood siding would be much more tolerant than unreinforced brick buildings.

Use of the Structure: Even small cracks in a house might be considered unacceptable, whereas much larger cracks in an industrial building may not be noticeable.

Presence of Sensitive Finishes: Tile or other sensitive finishes are much less tolerant of movements.

Rigidity of the Structure: If a footing beneath part of a very rigid structure settles more than the others then the structure will transfer some of the load away from the footing. However, the footings beneath flexible structures must settle much more before any significant load transfer occurs. Therefore, a rigid structure will have less differential settlement than a flexible one.

Aesthetic and Serviceability Requirements:The allowable settlement for most of the structures, especially buildings, will be governed by aesthetic and serviceability requirements and not by structural requirements. Unsightly cracks, jamming doors and windows and other similar problems will develop long before the integrity of the structure is in danger.

Settlement in building.

Table below shows the allowable foundation displacement into three divisions: Total settlement, tilting and differential settlement. It indicates that those structures that are more flexible or have more rigid foundations can sustain the larger values of total settlement and differential movement:

Type of Settlement	Limiting factor	Maximum Settlement
Total settlement	Drainage	15 – 30 cm
	Access	30 – 60 cm

	Probability of non-uniform settlement:	
	1. Framed structures	5 – 10 cm
	2. Masonry walled structures	2.5 – 5 cm
	3. Chimneys, silos, mats	8 – 30 cm
Tilting	Crane rails	0.003L
	Rolling of trucks etc.	0.01L
	Tilting of chimneys, towers	0.004L
	Stacking of goods	0.01L
	Drainage of floors	0.01 – 0.02 L
	Stability against overturning	Depends on H and L
Differential settlement	High continuous brick walls	0.0005 – 0.001 L
	Plaster cracking	0.001 L
	One-storey brick mill building, wall cracking	0.001 – 0.002 L
	Reinforced concrete building frame	0.0025 – 0.004 L
	Simple steel frame	0.005 L
	Steel frame, continuous	0.002 L
	Reinforced concrete building curtain walls	0.003 L

Where,

L = Distance between adjacent columns that settle to different amounts or between two points that settle differently.

The higher values are for regular settlements and more tolerant structures, whereas the lower values are for irregular settlement and critical structures,

H = Height.

W = Width of structure.

2.13 Codal Provision

Important Codal Provisions

The reference should be made to IS 3370 (Pan I and Pan II) 2009 and IS 33700 an IV/1%7 for design of RCC water tanks.

The code specify some COAMI:13111 over concrete mix to make it more impermeable and free from shrinkage cracks. The major provisions 101 concrete in R.C.C. water tank are given below: I.

Minimum grade of concrete should he M 30. It can he reduce hI NI 25 W the capacity of tank is up to 50m. The maximum cement content excluding fly ash should be 400kg. Maximum Water-Cement ratio should be less than 0.45.

Permissible Stress in Concrete

The permissible Stress in the concrete in tension will require to scenarist that concrete is safe against cracking. The values 101 permissible stress in concrete in direct tension and the flexural tension are given in table.

Considering the possible redistribution of flexural member, higher values of the permissible stress are specified for flexural tension. The rakes for the permissible stress in direct and flexural tension in concrete.

Permissible Stress in Steel

The tensile stress in the steel will necessarily be limited by the requirement that the permissible tensile stress in the concrete is not exceeded. So the tensile stress in steel will be equal to the product of modular ratio of steel and concrete and the corresponding permissible tensile stress in concrete. The value of the modular ratio for different grade of concrete is tabulated.

Permissible concrete stresses in calculations relating to resistance to cracking:

Sl.no	Grade of concrete	Permissible stress in direct tension (N/mm²)	Permissible stress in flexural tension (N/mm²)
1	M 25	1.3	1.8
2	M 30	1.5	2.0

Permissible concrete stress in compression and bond:

Sl. no	Grade of concrete	Permissible stress in flexural compression (N/mm²)	Permissible stress in direct compression (N/mm²)	Permissible bond stress in tension for plain bar (N/mm²)
1	M 25	8.5	6.0	0.9
2	M 30	10.0	8.0	1.0

Modular ratio for different grades of concrete:

Grade of concrete	Modular ratio
M 25	10.98
M 30	9.33

Minimum Reinforcement

If the dimension of the tank does not exceed 15m then minimum reinforcement is restricted to 0.24% in each direction and 0.35% for tanks with dimension exceeding 15m.

If the tank thickness exceeds 500mm then minimum percent of steel should be calculated for cross section area with thickness of 500mm only.

The calculated reinforcement area can be divided on both face of the tank wall. The maximum spacing of the reinforcement should not exceed the thickness of the member or 300mm, whichever is less.

In raft slab, for raft thickness less than 300, the minimum reinforcement should be placed on water face only. For higher thickness of raft slab, the minimum reinforcement is governed by IS 3370:2009 clause 8.1

2.14 Methods of Minimizing Total and Differential Settlements

Method	Comment
Procedures for linear fills on swamps or compressible surface stratum	
Excavation of soft material	When compressible foundation soils extends to a depth of about 10 to 15 ft, it may be practicable to remove entirely. Partial removal is combined with various methods of displacing remaining soft material.
Displacement by weight of the fill	Complete displacement is obtained only when the compressible foundation is thin and very soft. Weight displacement is combined with excavation of the shallow material.
Jetting to facilitate displacement	For a sand or gravel fill, jetting within the fill decreases its rigidity and promotes shear failure to displace soft foundation. Jetting within the soft foundation weakens it to assist itself in the displacement.
Blasting by trench or shooting methods	Charge is placed directly in front of advancing fill to blast out the trench into which the fill is forced by the weight of surcharge built up at its point. Limited to depths not exceeding about 20 ft.
Blasting by relief method	Used for building up fill on an old roadway or for fills of plastic soil. Trenches ale blasted at both toes of the fill slopes, relieving confining pressure and allowing the fill to settle and displace underlying soft materials.
Blasting by under fill method	Charge is placed in soft soil underlying the fill by jetting through the fill at a preliminary stage of its buildup. Blasting loosens compressible material, accelerating settlement and facilitating displacement to both sides. In some cases, relief ditches are cut or blasted at toe of the till slopes. Procedure is used to swamp deposits up to 30 ft thick.
Procedures for pre-consolidation of soft foundations	
Surcharge fill	Used where compressible stratum is relatively thin and sufficient time is available for consolidation under the surcharge load. Surcharge material may be placed as a stockpile for later use in permanent construction. Soft foundation should be stable against shear failure under surcharge load.

Accelerating consolidation by vertical drains	It is used where the tolerable settlement of the completed structure is small, where time available for pre-consolidation is limited and surcharge fill is reasonably economical. Soft foundation must be stable against shear failure under surcharge load.
Vertical drains with or without surcharge fill	Used to accelerate the time for consolidation by providing shorter drainage paths.
Well points placed in vertical sand drains	Used to accelerate consolidation by reducing the water head, thereby allowing increased flow into the sand drains. Particularly useful where potential instability of the soft foundation restricts placing of surcharge or where surcharge is not economic.
Vacuum method	Variation of well point in vertical sand drain but with a positive seal at the top of the sand drain surrounding the well point pipe. The atmospheric pressure replaces surcharge in consolidating soft foundations.
Balancing load of structure by excavation	Used in connection with mat or raft foundations on compressible material or where separate spread footings are found in suitable bearing material overlying compressible stratum. Use of this method may eliminate deep foundations, but it requires very thorough analysis of heave and soil compressibility.

Tolerance of Building

In most instances of construction, the subsoil is not homogeneous and also the load carried by various shallow foundations of a given structure will vary widely. Because a result, it is reasonable to expect varying degrees of settlement in various parts of a given building. The differential settlement of the parts of a building will lead to damage of the superstructure. Hence, it is vital to define certain parameters that quantify differential settlement and to develop limiting values for those parameters so as that the resulting structures be safe. Borland and Worth (1970) summarized the important parameters relating to differential settlement.

Figure shows a structure, in which different foundations, at A, B, C, 1 and E, have gone through some settlement. The settlement at A is AA', at B is BB', etc. Based on this figure, the definitions of the different parameters are as follows:

S_T = Total settlement of a given point.

ΔS_T = Difference in total settlement between any two points.

α = Gradient between two successive points.

β = Angular distortion = $\Delta ST_{(ij)} / I_{ij}$.

ω = Tilt.

Δ = Relative deflection (i.e. movement from a straight line joining two reference points).

Δ/L = Deflection ratio.

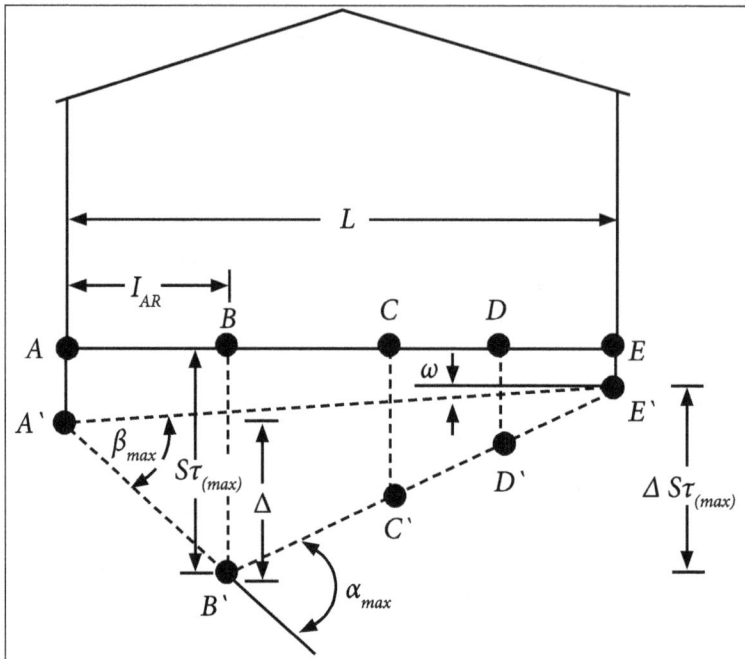

Problems

1. A clay layer 4 m thick is subjected to a pressure of 55 kN/m². If the layer has a double drainage and undergoes 50% consolidation in one year let us determine the co-efficient of consolidation. Take the time factor as 0.196. If the coefficient of permeability is 0.20 m/yr. Also determine the settlement in one year.

Solution:

Given:

> $H = 4$ m
>
> $\sigma = 55$ kN / m²

Double drainage d = H/2,

> $T_v = 0.196$
>
> $U = 50\%$
>
> $K = 0.02$ m / yr
>
> $t = 1$

To find: Co-efficient of consolidation $\rho_f = ?$

Settlement in one year.

Formula to be used:

$$T_v = \frac{C_v t}{d^2}$$

$$m_v = \frac{K}{C_v \cdot \gamma_\omega}$$

Final settlement $\rho_f = m_v \cdot \Delta\sigma' \, H$,

$$U = \frac{\rho}{\rho_f}$$

$$T_v = \frac{C_v t}{d^2} \qquad \therefore C_v = \frac{T_v d^2}{t}$$

$$C_v = \frac{0.196 \times 4/2}{1}$$

$$C_v = 0.392 \text{ m/yr}$$

$$m_v = \frac{K}{C_v \cdot \gamma_\omega}$$

$$= \frac{0.02}{0.392 \times 9.81}$$

$$m_v = 5.2 \times 10^{-3} \, \text{m}^2 / \text{kN}.$$

Final settlement $\rho_f = m_v \cdot \Delta\sigma' \, H$.

$$= 5.2 \times 10^{-3} \times 55 \times 4$$
$$= 1.14 \text{ m}.$$

Settlement after one year,

$$U = \frac{\rho}{\rho_f}$$

$$\therefore \rho = U \times \rho_f$$

$$= 0.4 \times 1.14$$

$$\rho = 0.57 \, \text{m}$$

2. A 6 m thick bed of clay is overlained by 9 m thick layer of sand with water table at 4 m below ground surface. The initial void ratio of the clay layer is 1.08 and compression index is 0.315. For the sand layer the bulk unit weight above and below the water table are 18 kN/m³ and 20 kN/m³ respectively. Let us calculate the settlement of a building constructed on sand layer if it causes an increase in effective vertical stress of 100 kN/m² at the middle of clay layer.

Solution:

Given:

        Thickness of clay bed = 6 m

        Thickness of sand = 9 m

Unit weight above and below water table are 18 kN/m³ and 20 kN/m³,

$$e_0 = 1.08; \; C_C = 0.315; \; \Delta\sigma' = 100 \; kN/m^2.$$

To find: Settlement of a building.

Formula to be used:

$$\gamma = \frac{G-1}{1+e} \gamma_w$$

$$\sigma'_o = 4 - \gamma_{sat} \; (sand \; above \; W.T) + 5 \times Y(sand) + 3 \times \gamma(clay)'$$

$$\Delta H = \frac{H}{1+e_o} C_c \cdot \log_{10} \frac{\sigma'_o + \Delta\sigma'}{\sigma'_o}$$

Assume specific gravity for clay as 2.70,

$$\gamma' = \frac{G-1}{1+e} \gamma_w$$

$$= \frac{2.7-1}{1+1.08} \times 9.81$$

$$\gamma' = 8.02 \; kN/m^3$$

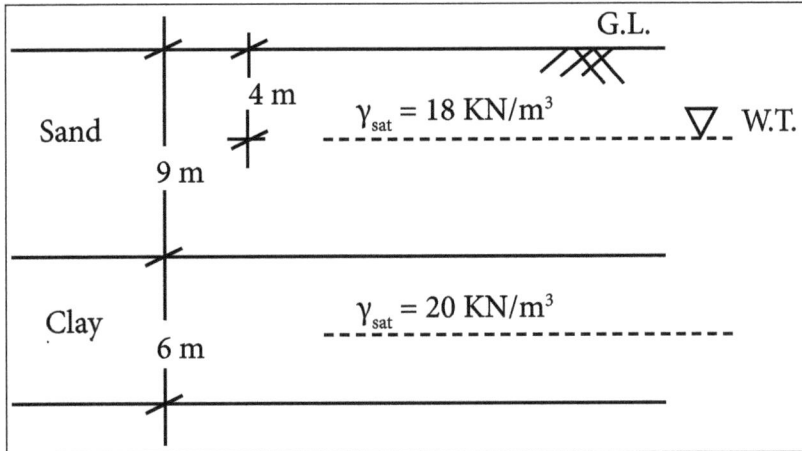

Initial pressure at the middle of clay layer.

(General equation $\sigma = \gamma Z$),

$$\sigma'_0 = 4 - \gamma_{sat} \ (\text{sand above W.T}) + 5 \times Y (\text{sand}) + 3 \times \gamma (\text{clay})'$$

$$= 4 \times 18 + 5 \ (20 - 9.81) + 3 \times 1.02$$

$$= 4 \times 18 + 5 \times 10.19 + 3 \times 8.02$$

$$= 72 + 50.95 + 24.06$$

$$= 147.01 \ kN / m^2$$

$$\Delta H = \frac{H}{1 + e_0} \ Cc \cdot \log_{10} \frac{\sigma'_0 + \Delta \sigma'}{\sigma'_0}$$

$$= \frac{6}{1 + 1.08} \ 0.315 \log_{10} \frac{147.01 + 100}{147.01}$$

$$H = 0.205 \ m$$

$$H = 205 \ mm.$$

Footings and Rafts

3.1　Types of Footings

Footing

A footing is a portion of the foundation of a structure that transmits loads directly to the soil.

Footings can be of the following types:

- Spread or isolated or pad footing.

- Strap footing.

- Combined footing.

- Strip or continuous footing.

- Mat or raft footing.

Spread Footing or Isolated or Pad Footing

It is circular, rectangular or square slab of uniform thickness. It is stepped to spread the load over a larger area. When footing is provided to support an individual column, then it is called "isolated footing".

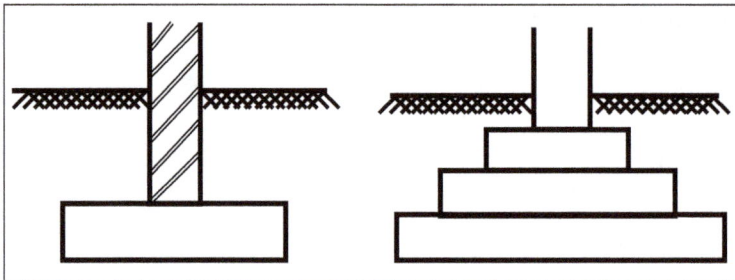

Isolated Footing.

Strap Footing

It consists of two isolated footings connected with a structural strap or a lever, as shown in figure below. Strap connects the footing such that they behave as one unit. Strap simply acts as a connecting beam. When the allowable soil pressure is relatively high and distances between the columns are large the strap footing is more economical than combined footing.

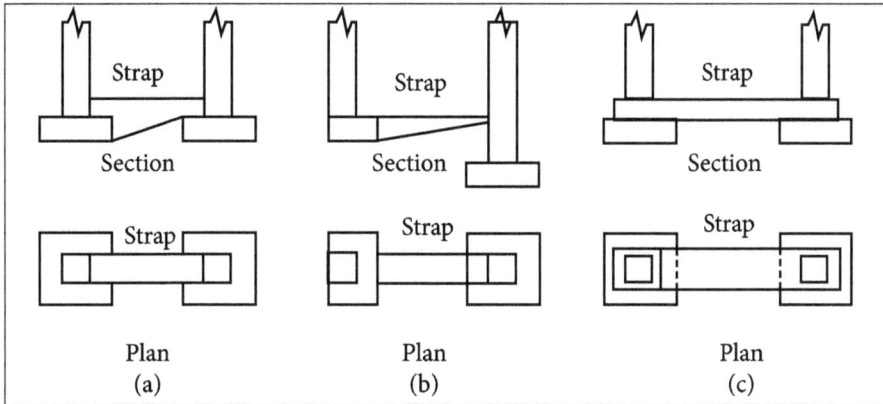

Strap footing.

Combined Footing

The combined footing supports two columns as shown in figure given below. It is used when the two columns are so close to each other that their individual footings would overlap, provided when the property line is so close to one column that a spread footing will be eccentrically loaded when kept entirely within the property line. By combining it with that of an interior column, the load is distributed evenly. A combine footing may be rectangular or trapezoidal in plan. The trapezoidal footing is provided when the load on one of the column is larger than the other column.

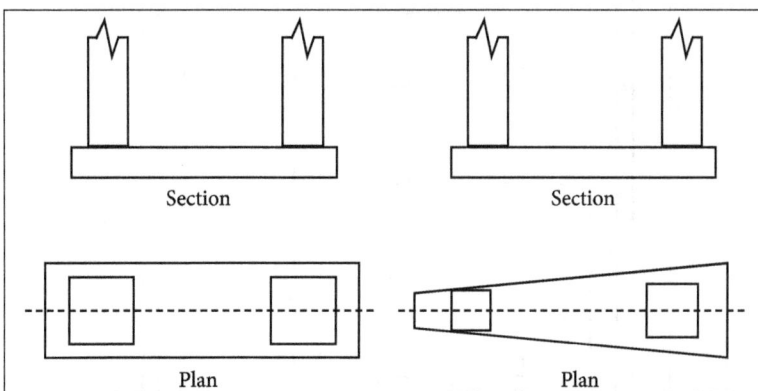

(a) Rectangular combined footing (b) Trapezoidal combined footing.

Strip Footing or Continuous Footing

It is another type of spread footing which is provided for a load bearing wall. It can also be provided for a row of columns which are so closely spaced that their spread footings overlap or nearly touch each other. In such a case, it is more economical to provide a strip footing than to provide a number of spread footings in one line. This is also referred as continuous footing.

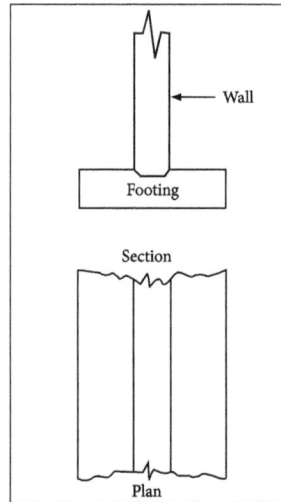

Continuous footing or strip footing.

Mat or Raft Footing

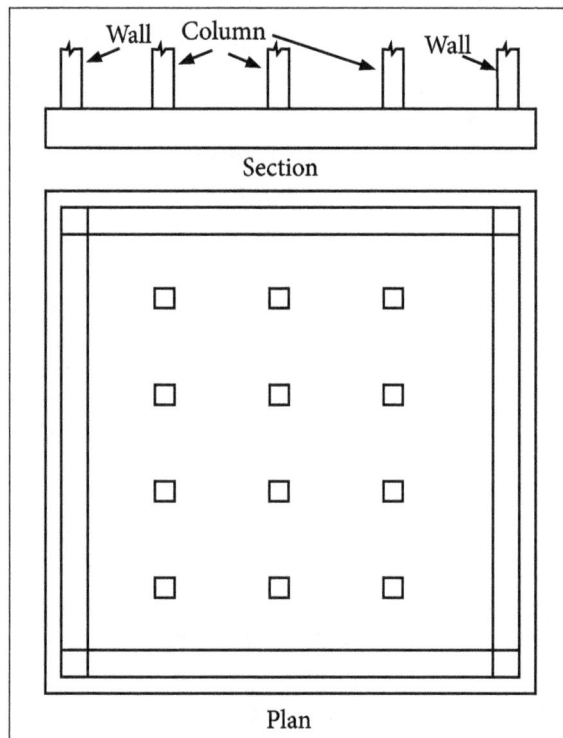

Mat footing.

It is a large slab supporting a number of walls and columns under entire structure or a large part of the structure. A mat is required when the allowable soil pressure is low or where the walls and columns are so close that individual footings would overlap or nearly touch each other. The mat foundations are useful in reducing the differential settlements on non-homogeneous soils or where there is a large variation in the loads on the individual columns.

Problems

1. A trapezoidal footing is to be produced which is to support two square columns of 30 cm and 50 cm sides respectively. Columns are 6 meters apart and the safe bearing capacity of the soil is 400 kN/m². The bigger column carries 5000 kN and the smaller 3000 kN. Let us design a suitable size of the footing so that it does not extend beyond the faces of the columns.

Solution:

Given:

Two square columns of 30 cm and 50 cm sides respectively.

Distance between columns = 6m.

Safe bearing capacity of the soil is 400 kN/m².

Formula to be used:

Area $A = \dfrac{a+b}{2} L = \dfrac{P_1 + P_2}{q_s}$

$$\therefore a+b = \dfrac{2}{6.8} \times \dfrac{5000+3000}{400} = 5.882\,m \qquad ...(1)$$

Also,

$$\bar{x} = \dfrac{0.5}{2} \times x' = 0.25 + \dfrac{3000 \times 6.4}{5000+3000} = 2.65\ m$$

But,

$$\bar{x} = \dfrac{L}{3}\left(\dfrac{2a+b}{a+b}\right) \qquad \therefore \dfrac{2a+b}{a+b} = \dfrac{3}{6.8} \times 2.65 = 1.169$$

or,

$$0.831\,a - 0.169b = 0 \qquad ...(2)$$

or,

$$b - 4.917\,a$$

Substituting this value in (1), we get $a = \dfrac{5.882}{5.917} = 0.994$ m and $b = 4.889$ m

Hence use trapezoidal footing of size a = 1 m; & b = 4.9m and L = 6.4 m

2. Design the proportion a strap footing for allowable pressures: 150 kN/m² for DL + reduced LL, 225 kN/m² for DL + LL.

Column loads	Column A	Column B
DL	500 kN	600 kN
LL	450 kN	800 kN

Let us proportion the footing for uniform pressure under DL + reduced LL. Distance of C/C of column =5.4 m; also here we design the projection beyond column A not to exceed 0.5 m.

Solution:

Given:

Allowable pressures: 150 kN/m² for DL + reduced LL, 225 kN/m² for DL + LL.

Distance of C/C of column =5.4 m

Column loads	Column A	Column B
DL	500 kN	600 kN
LL	450 kN	800 kN

Formula to be used: Column A and B: D.L. + Reduced L.L,

$$R_1 = \frac{P1 L}{L^1}$$

$$R_1 + R_2 = P_1 + P_2$$

$$L_1 = 2\left(e + \frac{C_r}{2}\right)$$

$$B = \frac{R_1}{L_1 q_s}$$

$$L_2 = \frac{P_2}{B q_s}$$

$$_{s1} \quad \overline{L \ B}$$

$$q_{s2} = \frac{R_2}{L_2 B}$$

Column A: D.L. + Reduced L.L. $= 500 + 0.5 \times 450 = 725 \text{ kN} = P_1$

Column B: D.L. + Reduced L.L. $= 600 + 0.5 \times 800 = 1000 \text{ kN} = P_2$

Basic equations to be used are,

$$R_1 = \frac{P_1 L}{L^1} \qquad ...(1)$$

$$R_1 + R_2 = P_1 + P_2 \qquad ...(2)$$

$$L_1 = 2\left(e + \frac{C_r}{2}\right) \qquad ...(3)$$

Here,

$$\frac{C_1}{2} = 0.5 \text{ m (given)}; \quad L = 5.4 \text{ m}; \qquad q_s = 150 \text{ kN/m}^2$$

Step 1:

Let e = 0.5 m. Hence L_1 = 2 (0.5 + 0.5) = 2 m

$$L' = L - e = 5.4 - 0.5 = 4.9 \text{ m}; \quad R_1 = \frac{P_1 L}{L'} = \frac{725 \times 5.4}{4.9} \approx 799 \text{ kN}$$

$$B = \frac{R_1}{L_1 q_s} = \frac{799}{2 \times 150} = 2.66 \text{ m}; \quad L_2 = \frac{P_2}{B q_s} = \frac{1000}{2.66 \times 150} = 2.506 \text{ m}$$

Thus, L_1, L_2, L' and B are known corresponding to this value of e.

Step 2:

$$R_1 = \frac{P_1 L}{L'} = \frac{725 \times 5.4}{4.9} \approx 799 \text{ kN}$$

$$R_2 = (P_1 + P_2) - R_1 = 725 + 1000 - 799 = 926 \text{ kN}$$

$$\therefore q_{s1} = \frac{R_1}{L_1 B} = \frac{799}{2 \times 2.60} = 153.6 \text{ kN}/\text{m}^2 > 150 \text{ kN}/\text{m}^2$$

$$q_{s2} = \frac{R_2}{L_2 B} = \frac{926}{2.50 \times 2.66} = 138.9 \text{ kN}/\text{m}^2 < 150 \text{ kN}/\text{m}^2$$

However, since $q_{s1} > q_s$, L_1 has to be increased.

This can be done by increasing the value of e.

Step 3: Let the next value of e be 0.6 m,

$$\therefore L_1 = 2\,(0.5+0.6)=2.2\,\text{m};\ L'=5.4-0.6=4.8\,\text{m}$$

$$R_1 = \frac{P_1 L}{L'}=\frac{725\times5.4}{4.8}=815.6\,\text{kN}\ ;B=\frac{R_1}{L_1 q_s}=\frac{815.6}{2.2\times150}=2.47\,\text{m}$$

Keep B = 2.5 m,

$$L_2=\frac{P_2}{Bq_s}=\frac{1000}{2.5\times150}=2.67\,\text{m}\approx2.7\,\text{m}$$

Step 4:

$$R_1=\frac{P_1 L}{L'}=\frac{725\times5.4}{4.8}=815.6\ ;R_2=725+1000-815.6=909.4\,\text{kN}$$

$$\therefore q_{s1}=\frac{815.6}{2.2\times250}=148.29<150\ \text{and}\ q_{s2}=\frac{909.4}{2.7\times2.5}=134.7<150$$

Hence we have,

$$B=2.5\,\text{m};\ L_1=2.2\,\text{m};\ L_2=2.7\,\text{m}$$

3. A concrete strip footing rectangular in section is located at ground level and extends 1.2 m below the ground level. It carries uniformly distributed load of 15000 kg/m. The soil profile consists of homogeneous clay of 6 m thick overlying rock. The clay properties are as under: Saturated unit bulk weight = 1750 kg/m³, Shear strength (un drained) = 8500 kg/m², Compressibility = 1 x 10⁻⁴ m²/100 kg. Let us determine the width of footing for factor of safety F (ii) ultimate consolidation settlement for F = 2. Also let us assume bulk unit weight of concrete = 2500 kg/m³. Neglect the spread of load beneath the footing and any side cohesion on the foundation.

Solution:

Given:

Distributed load $=15000\,\text{kg}/\text{m.}$

Compressibility $=1\times10^{-4}\,\text{m}^2/100\,\text{kg.}$

$c=8500\,\text{kg}/\text{m}^3.$

$\gamma_{sat}=1750\,\text{kg}/\text{m}^3.$

$\gamma'=1750-1000=750\,\text{kg}/\text{m}^3.$

To find:

Width of footing for factor of safety F.

Ultimate consolidation settlement for F = 2.

Formula to be used,

$$q_s = \frac{q_{nf}}{F} + \bar{\sigma} = \frac{q_{nf}}{F} + \gamma'D$$

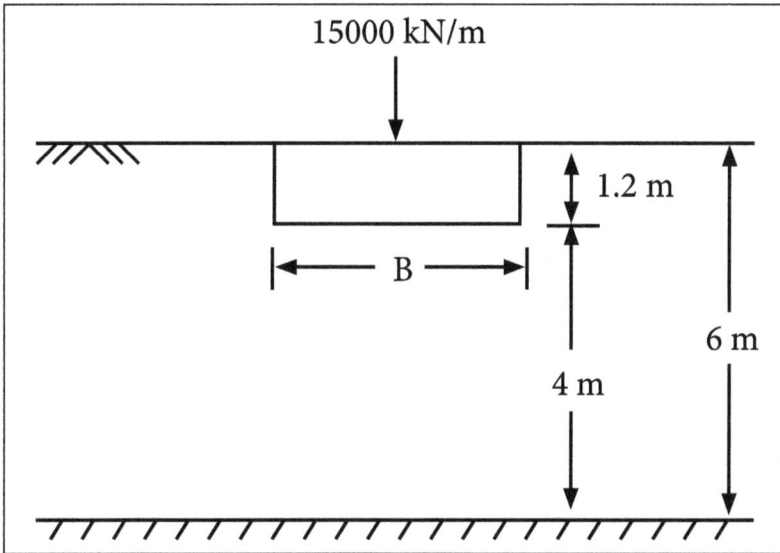

Width of Footing

$$q_{nf} = c\,N_c = 5.14\,c\,(\text{IS Code}) = 5.14 \times 8500 = 43690 \text{ kg}$$

Assuming water table at the ground level and using a factor of safety F on q_{nf}, safe bearing capacity is,

$$q_s = \frac{q_{nf}}{F} + \bar{\sigma} = \frac{q_{nf}}{F} + \gamma'\,D = \frac{43690}{F} + 750 \times 1.2 = \left(\frac{43690}{F} + 900\right)$$

Hence,

$$Q_s = \left(\frac{43690}{F} + 900\right) B \times 1 \text{ kg / m} \qquad ...(1)$$

Submerged unit weight of concrete $= 2500 - 1000 = 1500 \text{ kg / m}^3$,

$$q_a = 15000 \times 1 + B\,1.2 \times 1 \times 1500 = (15000 + 1800B) \text{ kg / m} \qquad ...(2)$$

Equating the two, we get,

$$15000 + 1800\,B = \left(\frac{43690}{F} + 900\right)B \quad \text{or} \quad B = \frac{15000}{\dfrac{43690}{F} - 900} = \frac{15000\,F}{43690 - 900\,F}$$

Which is the required value of B.

Taking,

$$F = 2$$

We get,

$$B = \frac{15000 \times 2}{43690 - 900 \times 2} \approx 0.72 \text{ m}$$

Ultimate Settlement for F = 2

Consider level AA at mid-depth below footing base. Given, coefficient of compressibility, $m_v = 1 \times 10^{-4} \text{ m}^2 / 100\text{kg} = 1 \times 10^{-6} \text{ m}^2 / \text{kg}$.

Load added by footing of width B = 0.72 m is,

$$= 15000 + 0.72 \times 1.2 \times 1500 - 0.72 \times 1.2 \times 750$$
$$= 15648 \text{ kg} / \text{m}.$$

Neglecting load spread,

$$\Delta \sigma' = \frac{15648}{B} = \frac{15648}{0.72} = 21733 \text{ kg/m}^2$$

$$\rho_f = M_v \Delta \sigma', H = 1 \times 10^{-6} \times 21733 \times 4.8 = 0.104 \text{m}.$$

4. A square footing is required to carry a net load of 1200 kN. Let us determine the size of the footing if the depth of foundation is 2m and tolerable settlement is 40 mm. The soil is sandy with N = 12, F.S. = 3, water table is very deep. Let us use Teng's equation.

Solution:

Given:

Load = 1200 kN

Depth of foundation = 2m

Tolerable settlement = 40 mm

N = 12

F.S. = 3

To find: Size of the footing.

Formula to be used,

$$q_p = 55(N-3)\left[\frac{B+0.3}{2B}\right]^2 R_{w2} \cdot R_d$$

$$R_d = 1 + \frac{0.2\,D}{B}$$

$$q_p = 55(N-3)\left[\frac{B+0.3}{2B}\right]^2 R_{w2} \cdot R_d$$

Here R_{w2} = 1 (since water table is very deep),

$$R_d = 1 + \frac{0.2\,D}{B} \le 1.2 = 1 + \frac{0.2 \times 2}{B} = 1 + \frac{0.4}{B}$$

$$\therefore q_p = 55(12-(3))\left[\frac{B+0.3}{2B}\right]^2\left(1+\frac{0.4}{B}\right) \qquad \ldots(1)$$

But,

$$q_a = \frac{1200}{B^2} \qquad \ldots(2)$$

$$\therefore \frac{1200}{B^2} = 55(12-3)\left[\frac{B+0.3}{2B}\right]^2\left(1+\frac{0.4}{B}\right) \qquad \ldots(3)$$

Assuming B ≤ 2 m, $R_d = 1 + \dfrac{0.4}{B} = 1.2,$

$$\therefore \frac{1200}{B^2} = 55 \times 9 \times 1.2\left[\frac{B+0.3}{2B}\right]^2 \text{ or } B^2\left[\frac{B+0.3}{2\,B}\right]^2 = 2.02$$

or,

$$(B + 0.3)^2 = 8,08 \text{ From which B} = 2.54 \text{ m}$$

Since B > 2 m, R_d will be less than 1.2. Hence from equation (3),

$$\frac{1200}{B^2} = 123.75\left(\frac{B+0.3}{B}\right)^2\left(1+\frac{0.4}{B}\right)$$

or,

$$B^2\left(\frac{B+0.3}{B}\right)^2\left(1+\frac{0.4}{B}\right)=9.7$$

or,

$$(B+0.3)^2\ (B+0.4)=9.7\ B$$

Solving this by trial and error, we get B ≈ 2.6 m.

5. A rectangular footing 2 m × 3 m carries a column load of 600 kN at a depth of 1 m. The footing rests on a c - ϕ soil strata 6 m thick, having Poisson's ratio of 0.25 and Young's modulus of elasticity as 20000 kN/m². Let us calculate the immediate elastic settlement of the footing.

Solution:

Given:

Column load = 600 kN

Depth = 1 m

Poisson's ratio = 0.25

Young's modulus of elasticity = 20000 kN/m²

To find: Immediate elastic settlement of the footing.

Formula to be used,

$$S_i = qB\left(\frac{1-\mu^2}{E_s}\right)I_w$$

Here,

q = intensity of contact pressure $=\dfrac{600}{2\times 3}=100\,\text{kN}/\text{m}^2$

B = least lateral dimension of footing = 2 m

μ = Poisson's ratio = 0.25; $E_s = 20000\ \text{kN}/\text{m}^2$

Iw = influence factor= 1.06 for rigid rectangular footing having $\dfrac{L}{B}=1.5$

$$\therefore S_i = 100\times 2\left[\frac{1-(0.25)^2}{20000}\right]\times 1.06 = 9.94\times 10^{-3}\,\text{m} = 9.94\,\text{mm}$$

6. A footing 3m × 1.5m in plan transmits a pressure of 160 kN/m² on a cohesive soil having $E = 8 \times 10^4$ kN/m² and $\mu = 0.48$. Let us determine the immediate settlement at the center assuming the footing to be (a) flexible and (b) Rigid.

Solution:

Given:

> Pressure = 160 kN/m²
>
> $E = 8 \times 10^4$ kN/m²
>
> $\mu = 0.48$

To find: Immediate settlement at the center.

L/B = 3/1.5 = 2. Hence we get I_w = 1.52 for flexible footing and 1.22 for rigid footing,

$$(a) S_i = 160 \times 1.5 \left[\frac{1 - 0.48^2}{8 \times 10^4} \right] \times 1.52 = 3.51 \text{ mm}$$

$$(b) S_i = 160 \times 1.5 \left[\frac{1 - 0.48^2}{8 \times 10^4} \right] \times 1.22 = 2.82 \text{ mm}$$

7. A square footing 1.2 m x 1.2 m rests at a depth of 1 m in a saturated clay layer 4 m deep. The clay is normally consolidated, having an unconfined compressive strength of 40 kN/m². The soil has a liquid limit of 30%, Υ_{sat} = 17.8 kN/m³, w = 28% and G = 2.68. Let us determine the load which the footing can carry safely with a factor of safety of 3 against shear. Also let us determine the settlement if the footing is loaded with this safe load. Use Terzaghi's analysis for bearing capacity.

Solution:

Given:

> Depth = 1 m.
>
> Depth of saturated clay layer = 4 m.
>
> Compressive strength = 40 kN/m².
>
> Liquid limit = 30%.
>
> $\Upsilon_{sat} = 17.8$ kN/m³.
>
> w = 28%.
>
> G = 2.68.

To find:

- Load.
- Settlement.

Formula to be used,

$$c = \frac{q_u}{2}$$

$$q_s = \frac{q_{nf}}{F} + \gamma_{sat} D$$

$$Q_s = q_s * B^2$$

$$S_c = C \frac{C_c}{1 + e_o} H \log_{10} \frac{\sigma_o + \Delta\sigma}{\sigma_o}$$

$$e_o = w_{sa} G$$

$$C_c = 0.009(w_L - 10)$$

Since,

$$= 0, N_c = 5.7; N_q = 1 \text{ and } N_y = 0. \text{ Also, } a = y_{sat} D = 17.8 \times 1 = 17.8$$

$$c = \frac{q_u}{2} = \frac{40}{2} = 20 \text{ kN/m}^2$$

$$q_s = \frac{q_{nf}}{F} + \gamma_{sat} D = \frac{1}{F}\left[1.3 c N_c + \overline{\sigma}(N_q - 1) + 0.04 B \cdot \gamma N_\gamma\right] + \gamma_{sat} D$$

or,

$$q_s = \frac{1}{3}\left[1.3 \times 20 \times 5.7 + 17.8(1-1) + 0\right] + 17.8 = 49.4 + 17.8 = 67.2 \text{ kN/m}^2.$$

$$Q_s = q_s * B^2 = 67.2 \times 1.2 \times 1.2 = 96.77 \text{ kN.}$$

Thickness of clay layer = 4 m. Depth of center of clay layer, below footing level $\frac{4}{2} - 1$.

Assuming load dispersion at 45° width of load spread= 1.2 + 2(1) = 3.2 m.

Vertical Stress Increment Due to Foundation Load

$$= \Delta \sigma = \frac{96.77}{3.2 \times 3.2} = 9.45 \text{ kN/m}^2$$

Now, the consolidation settlement of footing is given by,

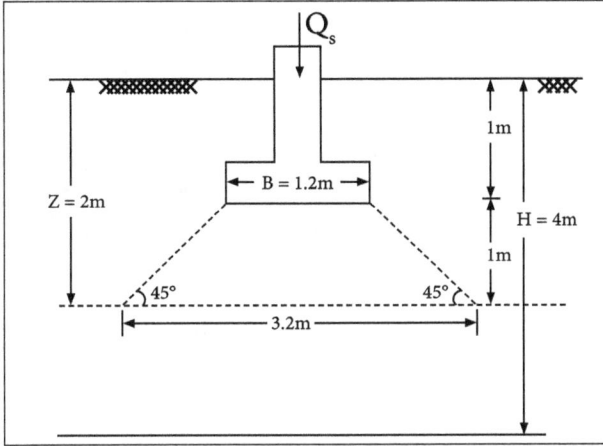

$$S_c = C \frac{C_c}{1+e_0} H \log_{10} \frac{\sigma_0 + \Delta\sigma}{\sigma_0}$$

Assume,

$$C = 1; e_0 = w_{sa} G = 0.28 \times 2.68 = 0.75$$

$$C_c = 0.009(w_L - 10) = 0.009(30 - 10) = 0.18$$

Initial overburden pressure at the center of clay layer $= \sigma_0 = y_{sat} \, z = 17.8 \times 2 = 35.6 \text{ kN/m}^2$

$$\therefore S_c = \frac{0.18}{1 \quad 0.75} \times 4 \log_{10} \frac{35.6 \quad 9.45}{35.6} = 0.042 \text{ m} = 42 \text{ mm}.$$

3.2   Contact Pressure Distribution

The shape of the deformation pattern varies depending on the flexibility of the foundation and type of soil. The figure depicts the relative distribution of soil contact pressures and displacements on cohesion less and cohesive soil. Linear contact pressure distributions from uniformly applied pressure q are generally assumed for settlement analysis. An applied load Q might cause an unequal linear soil contact pressure distribution.

UNIFORM PRESSURE

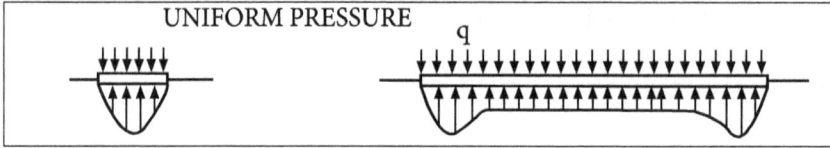

a. Rigid small footing on cohesionless soil b. Rigid mat on cohesive or cohesionless soil.

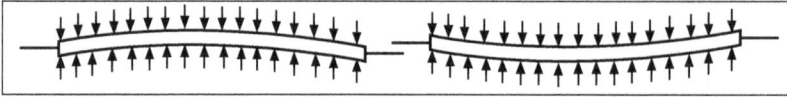

c. Flexible mat on cohesionless soil d. Flexible mat on cohesive soil.

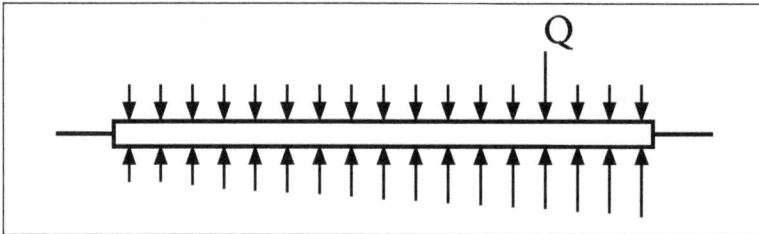

e. Linear contact pressure distribution.
Uniform pressure.

1. Cohesion Less Soil

It is composed of coarse-grained or granular materials with visually detectable particle sizes and also with little cohesion or adhesion between the particles. When unconfined these soils have little or no strength and when submerged they have little or no cohesion. Apparent adhesion between the particles in cohesion less soil can occur from capillary tension in pore water.

Settlement usually occurs quickly with little long-term consolidation and secondary compression or creep. Time rate effects may become significant in proportion to the silt content such that the silt content can dominate the consolidation characteristics:

- Uniformly loaded rigid foundations may cause less soil contact pressure near the edge than near the center, because this soil is pushed aside at the edges because of the reduced confining pressure. It leads to lower strength and lower modulus of elasticity in soil near the edge compared to the soil near the center. The parabolic soil contact pressure distribution can be replaced with a saddle-shaped distribution, for rigid footings or mats if the soil pressure does not approach the allowable bearing capacity.

- The distortion of a uniformly loaded flexible footing, mat or embankment on cohesion less soil will be concave downwards, since the soil near the center is stressed under higher confining pressure such that the modulus of elasticity of the soil is greater than near the edge.

- The theory of elasticity cannot be applied to cohesion less soil when the stress or loading increment varies significantly throughout the soil such that an equivalent elastic modulus cannot be assigned. The semi-empirical and numerical techniques have been very useful in determining the equivalent elastic parameters at points in the soil mass based on the stress levels that occur in the soil.

2. Cohesive Soil

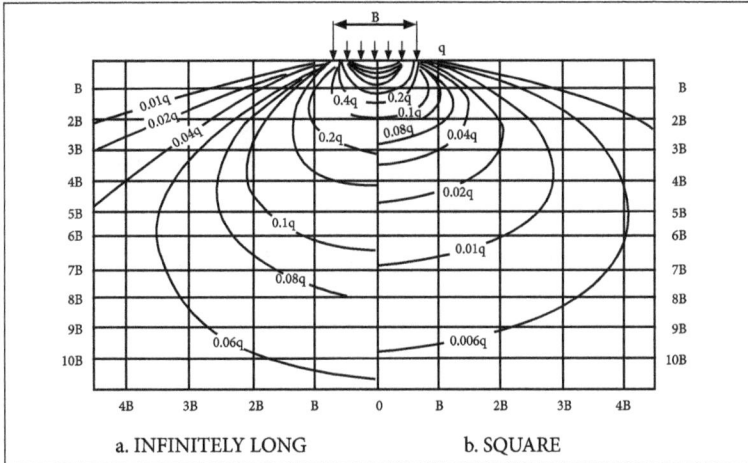

Cohesive soil.

The cohesive soil generally contains fine-grained materials consisting of silts, clays and organic material. When unconfined and air-dried these soils have significant strength. Most cohesive soil is relatively impermeable and when loaded deforms similar to a rubber or gelatin. i.e., the un drained state.

It includes granular materials with bonding agents between the particles such as soluble salts or clay aggregates. Wetting of soluble agents bonding granular particles can cause settlement in loose or high void ratio soil:

- A uniform pressure which is applied to a rigid foundation on cohesive soil.

Relative distribution of soil contact pressures and displacements of the rigid and flexible mats or footings on cohesion less and cohesive soils can cause the soil contact pressure to be maximum at the edge and minimum at the center, since additional contact pressure is generated to provide the stress that shears the soil around the perimeter:

- A uniform pressure which is applied to a flexible foundation on cohesive soil causes greater settlement near the center than near the edge because the cumulative stresses are greater near the center. As a result of the pressure bulb stress distribution indicated in Earth pressure measurements from load cells below a stiffening beam supporting a large, but flexible, ribbed mat also indicated large perimeter earth pressures which resembles a saddle-shaped pressure distribution.

- Elastic theory has been found useful in determining the immediate settlement when cohesive soil is subjected to moderate stress increments. The modulus of elasticity is a function of the shear strength of soil and often increases with increasing depth in proportion with the increase in shear strength of soil.

- Cohesive soil subject to stresses exceeding the maximum past pressure of the soil may settle down substantially from primary consolidation and secondary compression and creep.

Sources of Stress

Sources of stress in soil can occur from surface loads, soil weight and environmental factors such as desiccation from drought, changes in depth to groundwater and wetting from rainfall.

1. Soil Weight: Soil strata with different unit weights alter the stress distribution. Any change in the total stress results in changes in effective stress and pore pressure. In a saturated soil, any sudden rise in applied total stress results in a corresponding pore pressure increase.

This sudden increase may cause the flow of water out of the soil deposit, a decrease in pore pressure and an increase in effective stress. The changes in pore water pressure such as the raising or lowering of water tables also leads to a reduction or rise in effective stress.

2. Surface Loads: Loads applied to the surface of the soil mass increases the stress within the mass. The pressure bulb concept illustrates the change in vertical stress within the soil mass.

The placement of a uniform pressure over a foundation with a minimum width much greater than the depth of the soil layer would cause an increase of vertical stress in the soil which is approximately equal to the applied pressure.

3. Rules of Thumb for the Static Loads: Preliminary settlement analyses are sometimes performed before the structural engineer and the architect are able to furnish the design load conditions:

- The rules of thumb for line and column loads for buildings are based on a survey conducted by some engineering firms.
  - Tall multistory structures may have column loads exceeding 1000 tons.
  - Column spacing is often 20 to 25 ft or more.
  - The average pressure applied per story of a building often varies from 0.1 to 0.2 tsf.

- The vertical pressures from embankments may be estimated from the unit wet weight times height of the fill.

- The vertical pressures from dams, locks and retaining walls may be estimated by dividing the structure into vertical sections of constant height and determining the unit of weight times the height of each section.

Some Typical Loads on Building Foundations

Structure	Line load, tons/ft	Column load, tons
Apartment	0.5 to 1	30
Individual housing	0.5 to 1	< 5
Warehouses	1 to 2	50
Retail space	1 to 2	40
Two-storey building	1 to 2	40
Industrial facilities		50
Schools	1 to 3	50
Multi-storey building	2 to 5	100
Administration buildings	1 to 3	50

3.3   Isolated Footing

Design Guidelines

Specifications for Design of footings as per IS 456: 2000.

The important guidelines given in IS 456: 2000 for the isolated footing designs are given below:

- Footings are designed to sustain the applied loads, forces and moments and the induced reactions and to ensure that any settlement which may occur as uniformly as possible and the safe bearing capacity of the soil is not exceeded.

- In sloped or stepped footings the effective cross-section in compression is restricted by the area above the neutral plane.

- Angle of slope or depth and location of steps is provided such that the requirements of the design are satisfied at every section.

- Sloped and stepped footings that are designed as a unit can be constructed to assure action as a unit.

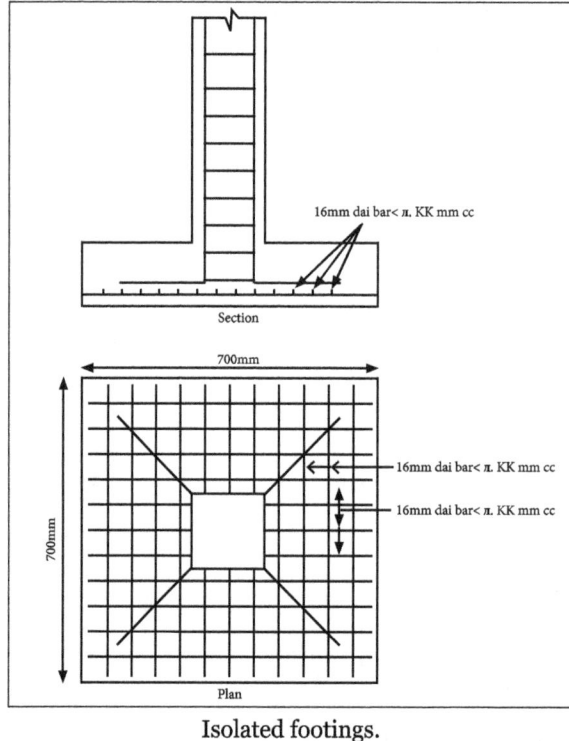

Isolated footings.

Thickness at the edge of footing:

- In plain and reinforced concrete footings, the thickness at the edge should not be less than 150 mm for footings on soils whereas, for footings on piles it should not be less than 300 mm above the top of the piles.

- In the case of plain concrete pedestals, the angle between the plane passing through the bottom edge of the pedestal and the corresponding junction edge of the column with the horizontal plane and pedestal shall be governed by the expression,

$$\tan \alpha < \neq 0.9 * \sqrt{(100 q_o / f_{ck}) + 1}$$

Where,

f_{ck} = Characteristic strength of concrete at 28 days in N/mm².

q_o = Calculated maximum bearing pressure at the base of the pedestal in N/mm².

Moments and Forces

In the case of footings on piles, computation for moments and shears may be based on the assumption that the reaction from any of the pile is concentrated at the center of the pile.

For the purpose of computing stresses in footings which supports a octagonal or round concrete column or pedestal, the face of the pedestal or column can be taken as the side of a square which is inscribed within the perimeter of a round or octagonal column or pedestal.

Bending Moment

It shall be determined by passing through a vertical plane which entirely extends across the footing and computing the moment of the forces acting over the entire area of the footing on one side of the said plane.

The greatest bending moment that is to be used in the design of an isolated concrete footing which supports a wall, column or pedestal shall be the moment computed in the manner as follows:

- At the face of the wall, column or pedestal for the footings supporting a concrete column, pedestal or wall.

- Halfway between the center-line and the edge of the wall for the footings under masonry walls.

- Halfway between the face of the pedestal or the column and also at the edge of the gusseted base, for footings under gusseted bases.

Shear and Bond

The shear strength of footings is governed by the following two conditions:

- The footing acting essentially as a wide beam, with a potential diagonal crack extending in a plane across the entire width, the critical section for this condition can be assumed as a vertical section located from the face of the wall, column or pedestal at a distance equal to the effective depth of the footing for footings on piles.

- Two-way action of the footing, with potential diagonal cracking along the surface of the pyramid or truncated cone around the concentrated load. In this case, the footing shall be designed for shear in accordance with the appropriate provisions.

In computing the external shear or any section through a footing supported on the piles, the entire reaction from any pile of diameter D_p whose center is located $D_p/2$ or more outside the section can be assumed as producing shear on the section.

The reaction from any pile whose center is located $D_p/2$ or more inside the section can be assumed as producing no shear on the section.

For intermediate positions of the pile center, the portion of the pile reaction should be assumed as producing shear on the section shall be based on the straight line interpolation between the full value at $D_p/2$ outside the section and zero value at $D_p/2$ inside the section.

The critical section to check the development length in a footing should be assumed at the same planes as those described for the bending moment and also at all the other vertical planes where abrupt changes of section occur.

If reinforcement is curtailed, the anchorage requirements must be checked in accordance with 26.2.3 of IS456: 2000.

Tensile Reinforcement

The total tensile reinforcement at any section will provide a moment of resistance at least equal to the bending moment on the section.

Total tensile reinforcement can be distributed across the corresponding resisting section as given below:

- In one-way reinforced footing, the reinforcement that extends in each direction should be evenly distributed across the full width of the footing.

- In two-way reinforced square footing, the reinforcement extending in each direction shall be evenly distributed across the full width of the footing.

- In two-way reinforced rectangular footing, the reinforcement in the long direction should be uniformly distributed across the full width of the footing.

For reinforcement in the short direction, a central band which is equal to the width of the footing shall be marked along the length of the footing and the portion of the reinforcement determined in accordance with the equation given below shall be uniformly distributed across the central band,

$$\frac{\text{Reinforcement in central band width}}{\text{Total reinforcement in short direction}} = \frac{2}{\beta + 1}$$

Where β is the ratio between the long side and the short side of the footing.

The remainder of the reinforcement should be distributed uniformly in the outer portions of the footing.

Transfer of Load at the Base of Column

The compressive stress in concrete at the base of a column or pedestal should be considered as being transferred by the bearing to the top of the supporting footing or pedestal. The bearing pressure on the loaded area shall not exceed the allowable bearing stress in direct compression multiplied by a value equal to $\frac{\sqrt{A_1}}{\sqrt{A_2}}$ but not greater than 2.

Where,

A_1 = Supporting area for bearing of footing.

A_2 = Loaded area at the column base.

When the permissible bearing stress on the concrete in the supporting or the supported member would be exceeded, the reinforcement shall be provided for developing the excess force, either by extending the longitudinal bars into the supporting member or by dowels.

When transfer of force is accomplished by reinforcement, the development length of the reinforcement shall be sufficient to transfer the tension or compression to the supporting member in accordance with 26.2 of IS456: 2000.

Extended longitudinal reinforcement or dowels of at least 0.5% of the cross-sectional area of the supported column or pedestal and a minimum of four bars shall be provided. When dowels are used, their diameter shall no exceed the diameter of the column bars by more than 3 mm.

Column bars of diameters greater than 36 mm, in compression only can be dowelled at the footings with bars of smaller size of the required area.

The dowel shall extend into the column, a distance which is equal to the development length of the column bar and into the footing and a distance equal to the development length of the dowel.

Nominal Reinforcement

The minimum reinforcement and the spacing should be as per the requirements of the solid slab.

The nominal reinforcement for the concrete sections of thickness greater than 1 m should be 360 mm² per meter length in each direction on each face.

This provision does not supersede the requirement of the minimum tensile reinforcement based on the depth of the section.

3.4   Combined Footings

The function of a foundation or a footing is to transmit the load from the structure to the underlying soil.

The choice of suitable type of footing mainly depends on the soil condition, the type of superstructure and the depth at which the bearing strata lie.

Definition

When two or more columns in a straight line are carried out on a single spread footing, it is known as combined footing.

Isolated footings for each column are generally economical.

Combined footings are provided only when it is absolutely needed, like:

- When two columns are close to each other, causing overlap of adjacent isolated footings.

- Proximity of building line or sewer or existing building, adjacent to a building column.

- Where soil bearing capacity is low, causing overlap of adjacent isolated footings.

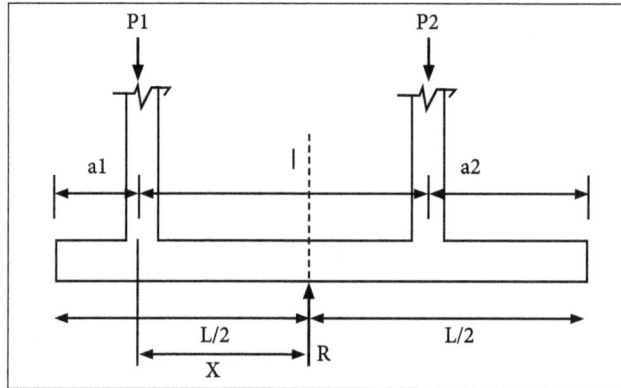
Combined footing with loads.

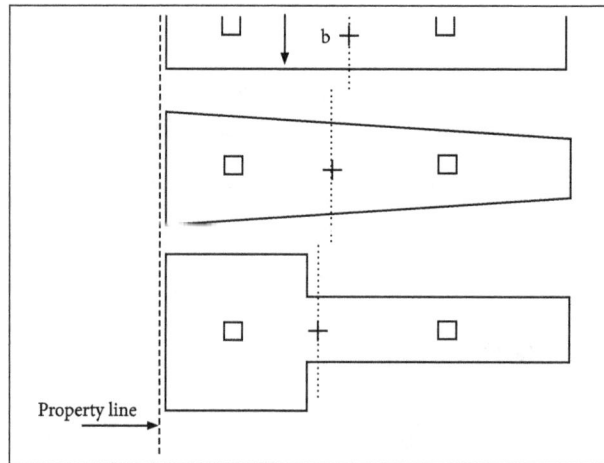
Combined footing.

Types of Combined Footing

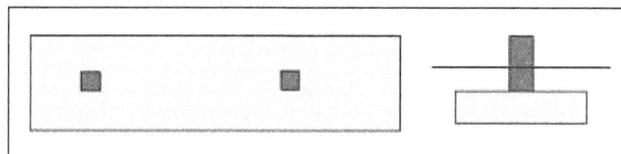
1. Slab type.

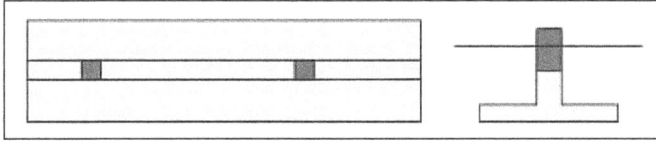
2. Slab and beam type.

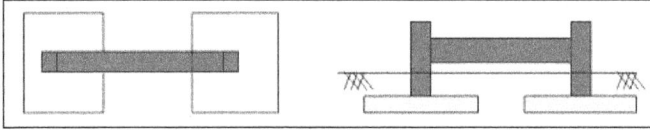
3. Strap type.

The combined footing may be rectangular, trapezoidal or tee-shaped in plan. The geometric shapes and proportions are so fixed that the centroid of the footing area coincides with the resultant of the column loads. This results in uniform pressure below the entire area of footing:

- Trapezoidal footing is provided when one column load is much greater than the other. As a result, both projections of footing beyond the faces of the columns will be restricted.

- The rectangular footing is provided when one of the projections of the footing is restricted or the width of the footing is restricted.

Design Steps

- Locate the point of application of the column loads on the footing.

- Proportion the footing such that the resultant of loads passes through the center of footing.

- Compute the area of footing such that the permissible soil pressure is not exceeded.

- Calculate the shear forces and bending moments at the salient points and hence draw the SFD and BMD.

- Fix the depth of footing from the maximum bending moment.

- Calculate the transverse bending moment and design the transverse section for depth and reinforcement. Check for anchorage and shear.

- Check the footing for longitudinal shear and hence design the longitudinal steel.

- Design the reinforcement for the longitudinal moment and place them in the appropriate positions.

- Check the development length for longitudinal steel.

- Curtail the longitudinal bars for economy.

- Draw and detail the reinforcement.

- Prepare the bar bending schedule.

Detailing

Detailing of steel in a combined footing is similar to that of a conventional beam. Detailing requirements of beams and slabs should be followed appropriately.

3.5   Types and Proportioning

Mat foundations are primarily shallow foundations. A brief overview of combined footings and the methods used to calculate their dimensions are as follows:

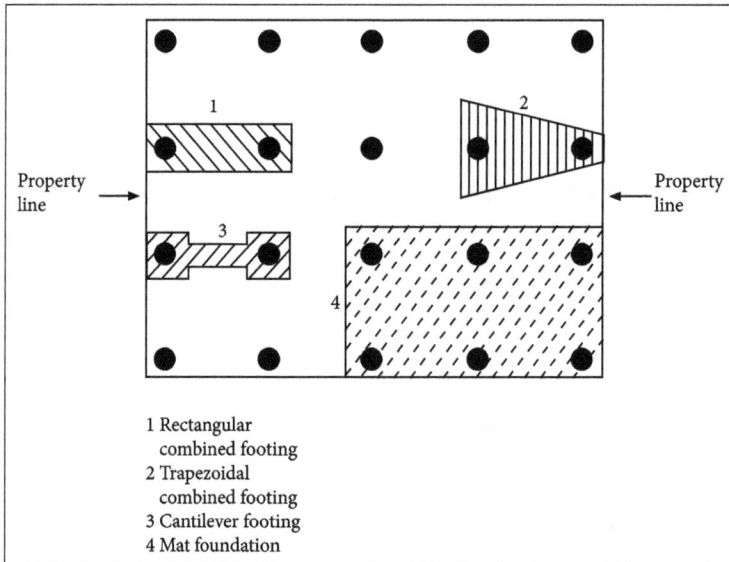

1 Rectangular
  combined footing
2 Trapezoidal
  combined footing
3 Cantilever footing
4 Mat foundation

(a) Combined footing.

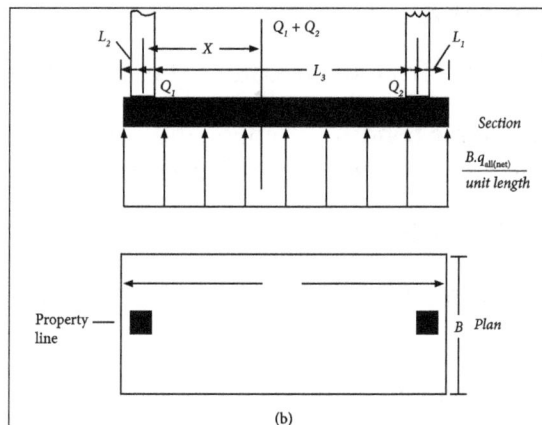

(b) Rectangular combined footing.

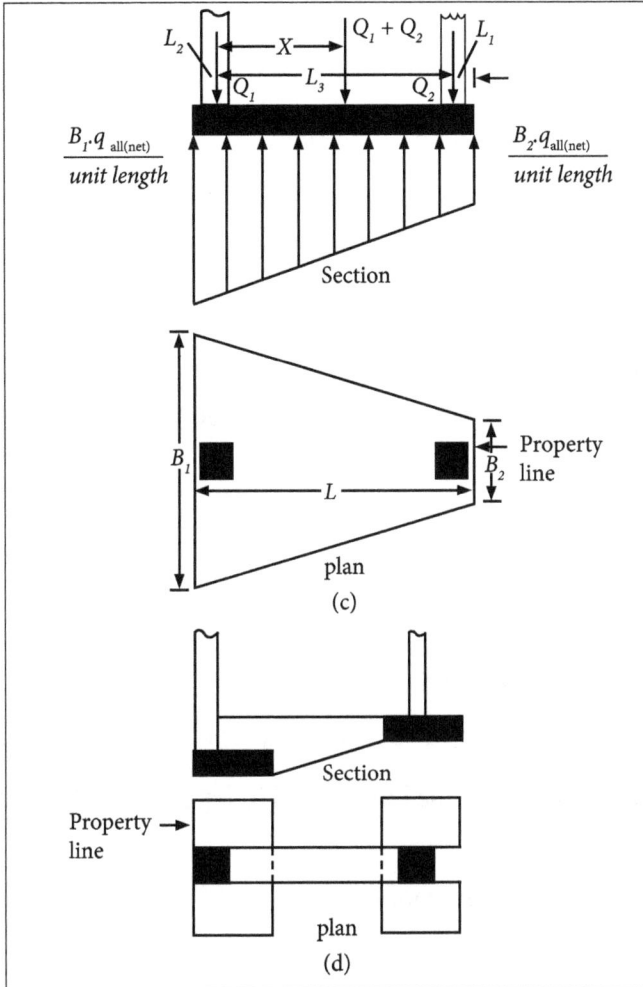

(c) Trapezoidal combined footing (d) Cantilever footing.

Rectangular Combined Footing

Load to be carried by a column and the soil bearing capacity are such that the standard spread footing design will require an extension of the column foundation beyond the property line.

In such a case, two or more columns may be supported on a single rectangular foundation, as shown in figure.

If the net allowable soil pressure is known, then the size of the foundation (B x L) can be determined in the following manner.

1. Determine the area of the foundation, A,

$$A = \frac{Q_1 + Q_2}{q_{all}\,(net)}$$

Where,

$Q_1 + Q_2$ = Column loads.

$q_{all\,(net)}$ = Net allowable soil bearing capacity.

2. Evaluate the location of the resultant of the column loads,

$$X = \frac{Q_2\,L_3}{Q_1 + Q_2}$$

3. For uniform distribution of soil pressure under the foundation, the resultant of the column load must pass through the centroid of the foundation.

Thus,

$$L = 2\left(L_2 + X\right)$$

Where,

L = Length of the foundation.

d. Once the length L is found out, the value of L_1 can be obtained: $L_1 = L - L_2 - L_3$.

Note that the magnitude of L_2 will be known and it depends on the location of the property line.

4. The width of the foundation then is,

B = A/L

Trapezoidal Combined Footings

Used as an isolated spread foundation of a column carrying a large load where space is tight.

Size of the foundation that will evenly distribute the pressure on the soil can be obtained in the following manner.

1. If the net allowable soil pressure is known, determine the area of the foundation,

$$A = \frac{Q_1 + Q_2}{q_{all(net)}}$$

From figure,

$$A = \frac{B_1 + B_2}{2} L$$

2. Determine the location of the resultant for the column loads,

$$X = \frac{Q_2 L_3}{Q_1 + Q_2}$$

3. From the property of Trapezoid,

$$X = L_2 = \left(\frac{B_1 + 2B_2}{B_1 + B_2}\right)\frac{L}{3}$$

Cantilever Footing

This type of footing construction uses a strap beam to connect an eccentrically loaded column foundation to the foundation of an interior column.

Used in place of trapezoidal or rectangular combined footings when the distance between the columns are large and the allowable soil bearing capacity are high.

Mat Foundation

It is a combined footing that may cover the entire area under a structure supporting several walls and columns.

It is sometimes preferred for soils that have low load-bearing capacities but that will have to support high column and wall loads.

Under certain conditions, spread footings would have to cover more than half of the building area and in such a case mat foundations might be more economical.

3.6   Mat Foundation

In case of soils having low bearing capacity, heavy structural loads are generally supported by providing raft or mat foundations. If the structure is at risk to subsidence on being located in the mining area or due to uncertain behaviour of its sub-soil water condition, the mat or raft foundations should be preferred.

The mat or raft foundations provides an economical solution to difficult site conditions, whereas pile foundation cannot be used advantageously and independent column footing becomes impracticable. The mat or raft foundations consists of thick reinforced concrete slab covering the entire area of the bottom of the structure like a floor.

Slab is reinforced with bars running at right angles to each other both near bottom and top face of the slab. It is necessary to carry the excessive column load by an arrangement of secondary beams and inverted main beams, cast monolithically with the raft slab.

The raft foundation is a thick concrete slab reinforced with the steel which covers the whole contact area of the structure like a thick floor. Area covered by raft might be greater than the contact area depending on the bearing capacity of the soil below. Reinforcing bars runs normal to each other in both top and bottom layers of the steel reinforcement.

Raft Foundation.

Method of Construction

- The raft slab generally projects for a distance of 30 cm to 45cm on all the sides of the outer walls of the structure as such the area of excavation is slightly more than the area of the structure itself.

- The excavation is made to the required depth and the entire excavated area is well consolidated.

- This surface, when dry, provides the base upon which the mat or raft slab is laid.

- Precautions that are necessary to be observed during the reinforced concrete construction should be strictly adhered.

- Further construction is started only after the curing of the raft has been fully done.

Problems

1. A building has to be supported on R.C. raft foundation of dimensions 14 m × 21 m. The soil is clay, which has an average unconfined compressive strength of 15 kN/m². The pressure on the soil due to the weight of the building and the loads that will carry will be 140 kN/m² at the base of the raft. The building has provision for basement floors. At what depth should the bottom of the raft be placed to provide a factor of safety of 3 against shear failure whose $\gamma_{clay} = 19$ kN / m³. Use Skeletons approach for bearing capacity calculations.

Solution:

Given:

Compressive strength $(q_u) = 15 \text{ kN} / \text{m}^2$.

Load $= 140 \text{ kN} / \text{m}^2$.

$\gamma_{clay} = 19 \text{ kN} / \text{m}^3$.

$c = q_u / 2 = 7.5 \text{kN} / \text{m}^2$; $B = 14$ m; $L = 21$m.

To find: Depth.

Formula to be used:

$$q_{nf} = 5c\left(1 + 0.2\frac{B}{L}\right)\left(1 + 0.2\frac{D}{B}\right)$$

$$q_s = \frac{q_{nf}}{F} + \gamma D$$

$$q_{nf} = 5c\left(1 + 0.2\frac{B}{L}\right)\left(1 + 0.2\frac{D}{B}\right) = 5 \times 7.5\left(1 + 0.2 \times \frac{14}{21}\right)\left(1 + 0.2\frac{D}{14}\right)$$

or,

$$q_{nf} = 42.5\left(1 + 0.01429\,D\right)$$

$$\therefore q_s = \frac{q_{nf}}{F} + \gamma D = \frac{42.5}{3}\left(1 + 0.01429D\right) + 19D \quad ...(1)$$

Actual load intensity,

$$q_a = 140 \text{ kN} / \text{m} \qquad ...(2)$$

Equating the two,

$$\frac{42.5}{3}\left(1 + 0.01429D\right) + 19D = 140$$

Or $19.2024\,D = 125.83$ from which $D = 6.55$ m.

2. A building to be supported on a reinforced concrete raft covering an area of 14×21 meters. The subsoil is clay with an unconfined compressive strength of 84 kN/m^2. The pressure on the soil due to weight of the building and loads it will carry, will be 120 kN/m^2, at

the base of the raft. If the unit weight of excavated soil is 15 kN/m³, let us calculate that at what depth should the bottom of the raft be placed to provide a factor of safety of 3.

Solution:

Given:

$$q_u = 84 \text{ kN}/\text{m}^2; \ c_u = q_u/2 = 42 \text{ kN}/\text{m}^2.$$

Load $= 120 \text{kN}/\text{m}^2$.

Weight of excavated soil is $15 \text{kN}/\text{m}^3$.

To find: Depth.

Formula to be used:

$$q_{nf} = 5c_u \left(1 + 0.2 \times \frac{B}{L}\right)\left(1 + 0.2\frac{D}{B}\right)$$

$$q_s = \frac{q_{nf}}{F} + \gamma D$$

Using Skemptons' equation for ultimate bearing capacity of raft,

$$q_{nf} = 5\,c_u \left(1 + 0.2 \times \frac{B}{L}\right)\left(1 + 0.2\frac{D}{B}\right) \qquad ...(1)$$

or,

$$q_{nf} = 5 \times 42 \left(1 + 0.2 \times \frac{14}{21}\right)\left(1 + 0.2\frac{D}{14}\right) = 238(1 + 0.01429D)$$

$$\therefore q_s = \frac{q_{nf}}{F} + \gamma D = \frac{238}{3}(1 + 0.01429D) + 15D = 79.33 + 16.134D \quad ...(2)$$

Actual load in density,

$$q_a = 120 \text{ kN}/\text{m}^2 \qquad ...(3)$$

Equating (2) and (3), we get 79.33 + 16.134 D = 120 from which D = 2.52 m

Check: Factor $5\left(1 + 0.2\dfrac{D}{B}\right) = 5\left(1 + 0.2 \times \dfrac{2.52}{14}\right) = 5.18 < 7.5$.

Hence OK.

3. A building is to be supported on a R. C. C. raft foundation of dimensions 12 m × 18m. The subsoil is clay which has an average unconfined compressive strength of 15 kN/m².

The pressure on the soil due to the weight of the building and the loads it will carry is expected to be 130 kN/m² at the base of the raft. If the unit weight of the excavated soil is 18 kN/m³, let us determine the depth at which the bottom of the raft should be placed to provide a factor of safety, of 3 against shear failure.

Solution:

Given:

Compressive strength = 15 kN/m².

Load = 130 kN/m².

Weight of the excavated soil is 18 kN/m³.

To find: Depth.

Since the building will have basement, the excavated soil is not going to be replaced.

Hence it is possible to allow an increase of yD in the safe loading intensity, i.e. + yD. Equating this value of q_s to the actual load intensity of 130 kN/m².

We get,

$$14.17 \left[1 + \frac{D}{60}\right] + 18D = 130$$

or,

$$0.245\,D + 18\,D = 130 - 14.17$$

From which D = 6.35 m.

3.7 Types and Applications

Types of Mat Foundation

The types of mat foundation generally used are:

- Flat plate mat.
- Flat plate thickened under columns.
- Two-way beam and slab.
- Plate with pedestal.
- Rigid frame mat.
- Piled raft.

Flat Plate Mat

A flat plate mat is used for fairly uniform and small column spacing and relatively light loads.

A flat plate type of mat is suitable when the soil is not too compressible.

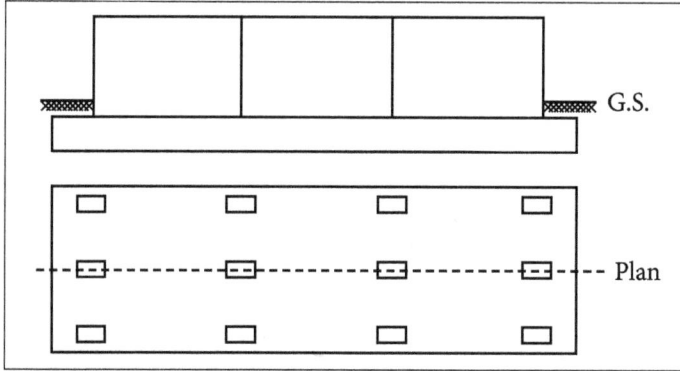

Flat Plate Mat.

Flat Plate Thickened under Columns

For columns subjected to very heavy loads generally the flat plate is thickened under columns as shown in figure to guard against diagonal shear and negative moments.

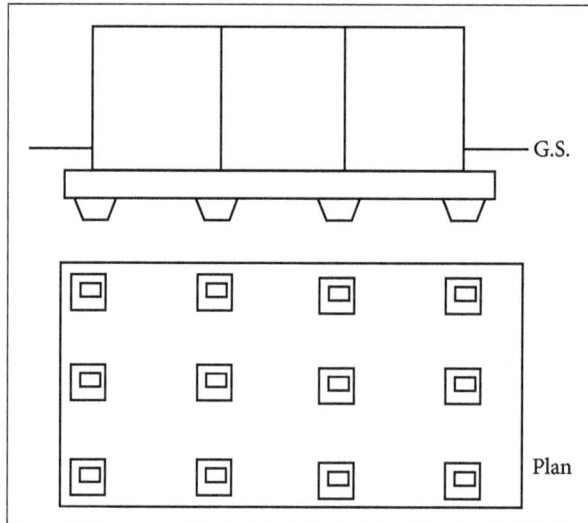

Flat Plate Thickened under Columns.

Two-way Beam and Slab

When the column spacing is large and carries unequal loads it will be more economical if a two-way beam and slab raft as shown in figure is used. This type of mat is generally suitable when underlying soil is too compressible.

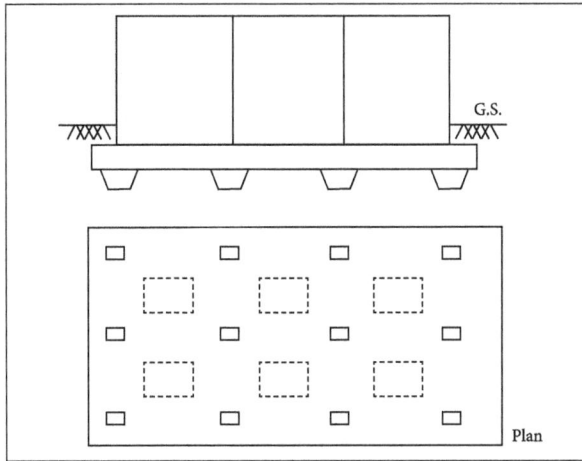

Two-way Beam and Slab.

Plates with Pedestals

The function of this mat is same as that of a flat plate thickened under columns. In this mat pedestals are provided at the base of the columns.

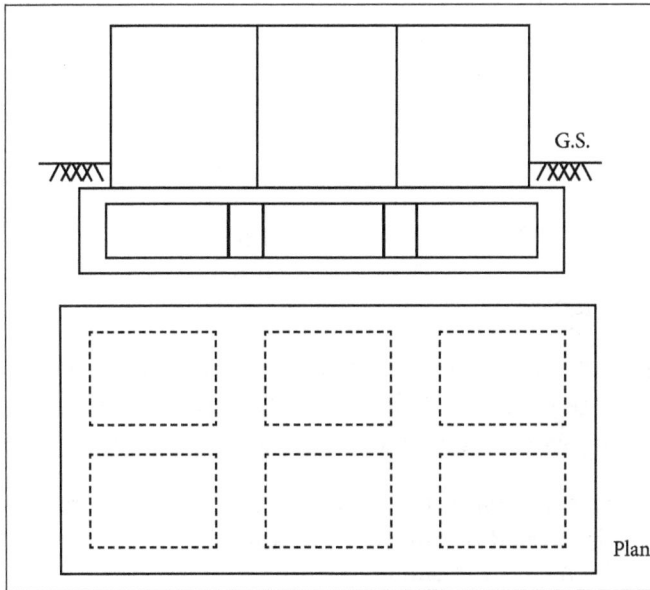

Plates with Pedestals.

Rigid Frame Mat

This type of mat is used when columns carry extremely heavy loads. In such a design, basement walls act as ribs or deep beam.

When the depth of beam exceeds 90 cm in simple beam and slab mat, a rigid frame mat is preferred.

Piled Raft

In this type of construction the mat is supported on piles. It is used where the soil is highly compressible and the water table is high. It reduces settlement and control buoyancy.

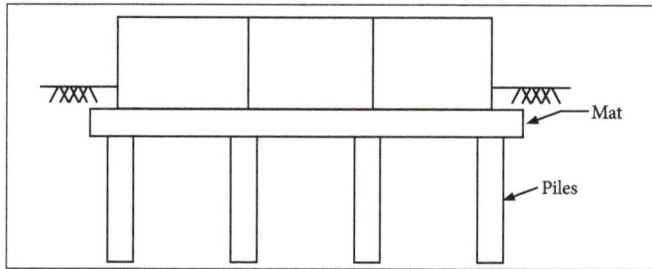

Piled Raft.

3.8   Proportioning

Foundations, whether continuous, as in a foundation wall or isolated, as when divided into piers, must have the footing courses proportioned to the weight they will be required to carry and to the bearing capacity of the soil.

The pressure on the soil from each square foot of the footings should be the same, where the soil is uniform and at no place should the bearing power of the soil be exceeded.

To secure the most satisfactory results, the footings must be proportioned to properly distribute the weight they are to carry over the sufficient areas of ground, to secure uniform settlement in each case.

If these conditions were properly considered, there would be a few cracks in the mason work, as such cracks are caused usually by unequal settlement. A uniform settlement even of an inch or more would in most of the buildings pass unnoticed.

In order to proportion the area of the footings, the weights coming on each pier and the weight of and loads carried by, all the walls should be computed and entered into a memorandum book for future reference.

The ground should be examined and by means of table - the bearing power of different soils, the load per square foot which the footings shall carry, can be determined.

The load on the various footings divided by this unit load, will give the proper area of each, in square feet.

The pressure under a brick pier that supports a tier of columns can be assumed at 10% less than the calculations, when the exterior support of the building is a brick wall, for the joints in the brickwork will close slightly under the weight and cause about 10 % more settlement than that will exist in the columns, each of which, being one piece, is practically joint less and hence will settle less.

One of the objects in proportioning the footings is to provide for uniform settlement in all parts of the building, so that the floors may remain level and that no cracks may occur in the walls.

Thus it will be seen that the dead load under the walls of a five-story building would be a considerable item whereas, the dead load under a tier of iron columns would be much less in proportion to the floor area supported.

As the dead load is always constant and the live load can vary greatly, only the amount of live load that will probably be supported by the footings should be considered.

For warehouses, stores, etc., 50 per cent of the live load that the floor beams have to carry should be added to the dead load carried on the footings.

For office dwelling houses, buildings, hotels, etc., the weight of the people occupying them need not enter into the calculations for proportions of footings and only from 25 to 80 pounds per square foot of floor will be allowed for the weight of furniture, books, safes, etc.

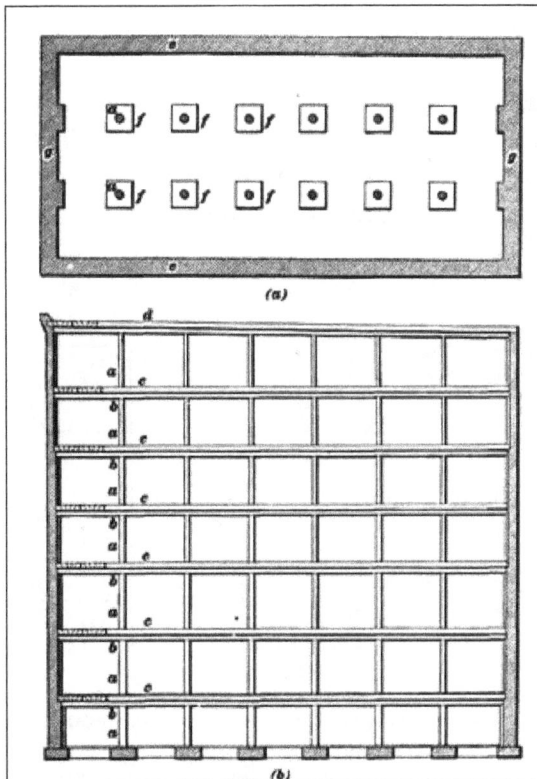

Proportioning of mat foundation.

3.9   Floating Foundation

It is defined as a foundation in which the weight of the building is approximately equal to the full weight including water of the soil removed from the site of the building. This flotation principle may be explained with reference to structure built in the excavation and completely filling it.

If the weight of the building is equal to the weight of the soil and water removed from the excavation, then it is evident that the total vertical pressure in the soil below depth D in figure (c) is same as that in figure (a) before excavation. Since the water level has not changed, the effective pressure and the neutral pressure are therefore remains unchanged. Since settlements are caused by an increase in effective vertical pressure, it would not settle at all.

This is the principle of a floating foundation, an exact balance of weight removed against the imposed weight. The result is zero settlement of the building. The excavation stage of the building is the most critical stage. Cases may arise where we cannot have a fully floating foundation.

The foundations of this type are sometimes known as partly compensated foundations. While dealing with floating foundations, we have to consider the following two types of soils.

Type 1

The foundation soils are of such a strength that the shear failure of soil will not occur under the building load but the settlements and especially differential settlements, will be too large and will constitute to the failure of the structure. Floating foundation is used to reduce settlements to an acceptable value.

Type 2

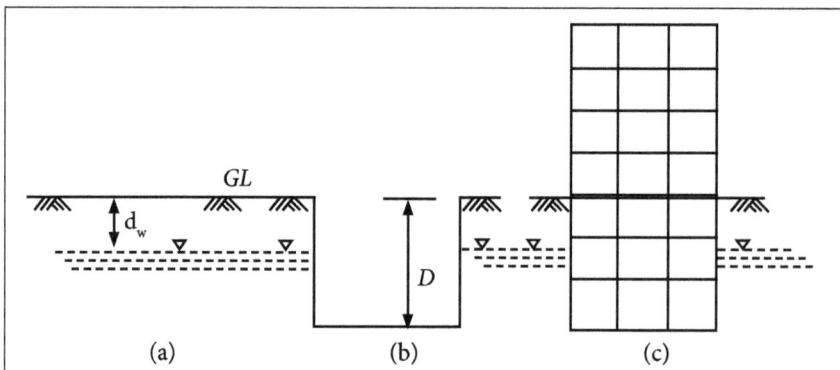

Balance of stresses in foundation excavation.

The shear strength of the foundation soil is too low that the rupture of the soil would occur if the building were to be founded at the ground level. In the absence of a strong layer at a considerable depth, the building can only be built on a floating foundation which decreases the shear stresses to an acceptable value. Solving this problem solves the problem of settlement.

In both the cases, a rigid raft or box type of foundation is required for the floating foundation.

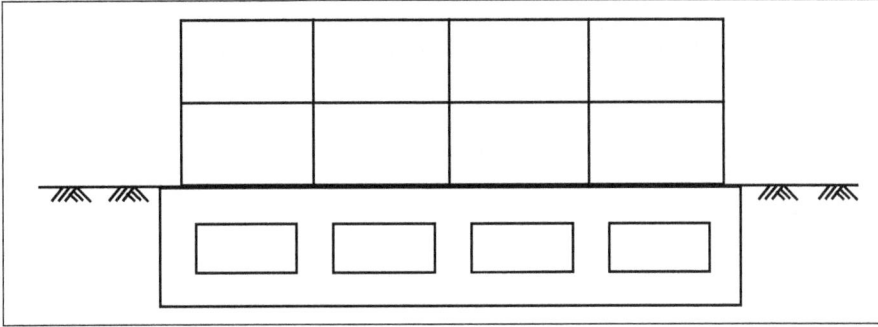

Rigid raft foundation.

3.10 Seismic Force Consideration

In seismic zones foundations are subjected to additional forces due to earthquakes. This will in turn depend on the area and intensity of the earthquake. Design should take into consideration these forces.

Foundations in seismic zones are subjected to additional forces because of the seismic waves. If the structure is not capable of withstanding these forces, it might collapse leading to serious catastrophes.

A foundation is the lowest part of a structure, which is generally below the ground level. Foundations are provided to transmit the load of the superstructure to the underlying soil.

It is necessary that the subsoil is capable of withstanding these loads. For this, the superimposed load must be lower than the soil's safe bearing capacity.

In case of foundations in seismic zones, additional loads are created because of the seismic vibrations. The design should consider all the additional forces.

General Considerations in the Seismic Design of Foundations

- Site investigations and determination of soil properties.

- Details of geological and geo technical environment.

- Identification of loads - static and dynamic.

- Type of foundations.

- Safety verification as per building codes.

Soil Investigations for Seismic Designs

For the design of foundations, the soil properties are to be ascertained.

For this, both static and dynamic tests must be to be conducted in the laboratory as well as in the field.

The important soil parameters to be analyzed are:

- Damping factors.

- Particle size distribution.

- Relative density.

- Shear modulus.

The important field tests to be conducted are:

- Cone Penetrometer Test.

- SPT Hammer Energy.

- Pressure Meter Testing.

- Shear wave velocity Measurement.

- Seismic Piezo cone Penetrometer.

The purpose of these laboratory and field tests are to define the hydraulic conditions, soil deposit details, soil index properties and static and dynamic stress-strain soil behaviour.

Main Factors that Influence Site Effects

Geotechnical:

- Nonlinear behaviour of soils.

- Elastic vibration characteristics of soils.

Seismological:

- Duration of bed rock motions.

- The intensity and frequency characteristics of the bed rocks in seismological environment.

Geometrical:

- Non-horizontal soil deposit layering.
- Topography of underlying bedrock.

Geological:

- Soil deposit thickness.
- Type of under lying rock.

Earthquake Characteristics

The horizontal forces that are generated in the structure is the main effect of an earthquake. The horizontal ground accelerations give a measure of this force. The granular soils get compacted because of the vibrations. This in turn causes the settlement of the ground surface.

Another effect of the vibration is known as liquefaction. The degree of liquefaction depends on the depth of water table, relative density of the soil, percentage of fines and the ground acceleration.

Effects of Earthquake

- The vibration due to earthquake can cause structural damage.
- The dynamic stress and induced pore water pressure can reduce the bearing capacity of the soil.
- Compaction of loose granular soil may induce excess pore water pressure, which in turn causes liquefaction of the soil.
- Loose granular soils are compacted by ground motion. This causes a large subsidence of the ground surface.

Measures to Prevent Liquefaction

- Compaction of loose soil:
  - With vibratory rollers.
  - Compaction piles.
  - Vibro flotation blasting.
- Grouting and chemical stabilization.
- Application of surcharge.

Conclusion

For the design of foundations in seismic regions, knowledge in a wide spectrum of disciplines such as soil dynamics, geology, seismology and structural dynamics is needed.

The design must be safe and at the same time economical.

3.11 Codal Provision

Indian standard code of practice for design and construction of raft foundation - IS 2950-1965.

There are two approaches for design. They are conventional method and the elastic method.

In the conventional method, the foundation is considered infinitely rigid and the pressure distribution is independent of the deflection of the raft. Soil pressures are also assumed to be planner such that the centroid of the soil pressure coincides with the line of action of the resulting forces of all the loads acting on the foundation. This method is generally used in design because of its simplicity.

A large amount of reinforcement is provided to safeguard the uncertainties caused by differential settlement. The raft is analyzed as a whole in each of the two perpendicular directions. Thus, the total shear forces and total bending moments acting on any section cutting across the entire raft is equal to the arithmetic sum of all the forces and reactions/moments to the right or left of the section. The actual reinforcement provided shall be twice that worked out theoretically. Elastic method has two approaches. In the first approach, the soil is replaced by an infinite number of isolated springs.

In the next approach, the soil is assumed to be a continuous elastic medium obeying the Hook's Law. These methods are applicable in case if the foundation is comparatively flexible and the loads tend to concentrate over the small areas. The actual reinforcement can be one-and-a-half times than that required theoretically. The famous method of soil line falls under this category.

The code mentions some of the codal provisions:

- Do not apply to large and heavy industrial construction where special considerations of the base pressure distribution will be required.

- Apply only for fairly horizontal planes of separation of layer and to fairly uniform soil conditions below.

- Foundations in seismic area and/or to vibrating load can be given special considerations. This code has been revised in 1973.

Raft Foundation

The design of soil line method was proposed by A.L.L. Baker. According to Mr. Baker, the design of raft as a reversed floor is dangerous.

Engineers being aware of this, therefore, generally adopt the second method in which the earth pressure is assumed to be uniform throughout and the moments are obtained at any section by statics.

In the second method also high values of moments are obtained, which may or may not be present and it is irrational or of waste to provide for such moments without investigating the variation and deflections in the soil pressure.

Therefore, Mr. Baker, suggested the soil line method which considers the variations in soil pressure and its relation to deflection but in order to minimize the calculations.

It is assumed that the earth pressure varies throughout the beam according to the law of straight line.

Indian Standard Code of Practice for Design and Construction of Raft Foundation IS: 2950 (Part-I) 1973

In the revised version of the code, the following methods of analysis have been proposed:

- Assumptions of linearly varying contact pressure.
- Perfectly flexible structures.
- Perfectly rigid structures.
- Structures stiffened along one axis.
- Structures stiffened along both the axis.
- General methods.

General methods are further classified based on:

- Modulus of subgrade reaction.
- Modulus of compressibility.

Method (a) corresponds to the conventional method of the code and has similar limitations.

Method (b) is applicable for structures which have relatively less stiffening members specially those resting on very stiff foundation soil. In this case, the deflections of the raft are the same as that of the settlements of the foundation soil under the external load.

In method (c), contact pressure distribution should be calculated based on the Boussineq's equation for Elastic Isotropic half space. It is applicable when deformations of raft under loads are small as compared to the mean settlement of the structure.

Method (d) is somewhere in between methods (b) and (c).

In the direction of the stiffened axis the contact pressure distribution is found out by Boussineq's equation as in method (c).

Method (e) is same as method (c). The two methods under method (f) are elastic methods and are used when simplified methods from (a) to (c) are not applicable. Details given in the code does not provide sufficient guidance to enable the analysis and design to be completed by the designer.

Apart from the limitations applicable in the code it is stated that:

- Allowable settlement both total and the differential settlement will satisfy the requirement of the super-structure.

- The approximate values of permissible settlements as given in earlier code have been deleted.

Pile Foundation

4.1   Types of Piles and their Function

Classification of pile with respect to load transmission and functional behavior:

- End bearing piles.

- Friction piles.

- Combination of friction and cohesion piles.

End Bearing Piles

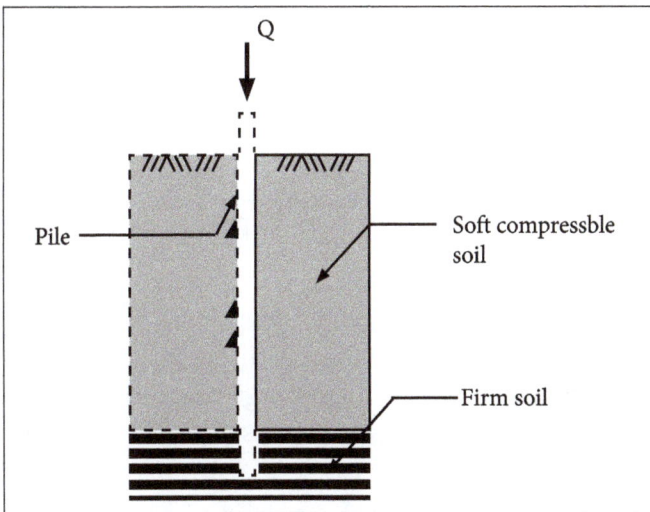

End bearing piles.

These piles are transferring their load on to a firm stratum located at a considerable depth below the base of the structure and they derive most of their carrying capacity of the penetration resistance of the soil at the toe of the pile. The pile acts as an ordinary column and should be designed in such a way. Even in a weak soil a pile will not fail by buckling and this effect will only be considered if part of the pile is unsupported, i.e. if it is in either air or water.

The load is transmitted to the soil through the friction or cohesion. But sometimes, the soil surrounding the pile will adhere to the surface of the pile and causes "Negative

Skin Friction" on the pile. Sometimes have a considerable effect on the capacity of the pile. Negative skin friction is caused by the consolidation of the soil and drainage of the ground water. The founding depth of the pile is influenced by the results of the site investigate and on soil test.

Friction or Cohesion Piles

Carrying capacity is derived from the adhesion or friction of the soil in contact with the shaft of the pile.

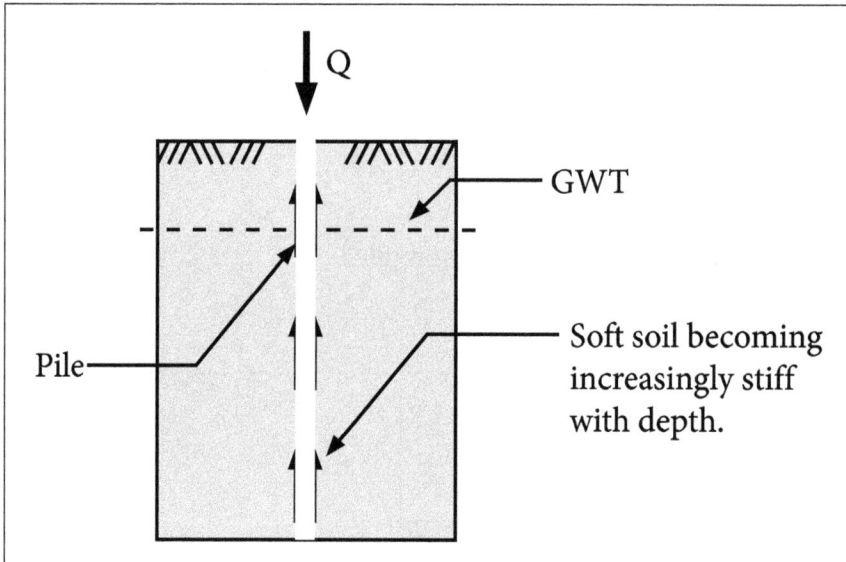

Friction or cohesion pile.

Cohesion Piles

These piles are transmitting their load to the soil through the skin friction. This process of driving such piles close to each other in groups greatly reduces the compressibility of the soil and porosity within and around the groups. Therefore piles of this category are sometimes known as compaction piles.

During the process of driving the pile into the ground, the soil becomes molded and as a result loses its strength. Therefore the pile is not able to transfer the exact amount of load which it is intended to do immediately after it has been driven. Usually the soil regains some of its strength after a period of three to five months since it has been driven.

Friction Piles

Friction piles also transfer their load to the ground through the skin friction. The process of driving such piles does not compact the soil appreciably. These types of pile foundations are commonly known as a floating pile foundations.

Combination of Cohesion Piles and Friction Piles

An extension of the end bearing pile occurs when the bearing stratum is not hard such as firm clay. The pile can be driven far enough into the lower material to develop sufficient frictional resistance.

A farther variation of the end bearing pile is the pile with enlarged bearing areas. This can be achieved by forcing a bulb of concrete into the soft stratum immediately above the firm layer to give an enlarged base.

A similar effect is produced with bored piles by forming a large bell or a cone at the bottom with a special reaming tool. The bored piles which are provided with a bell have high tensile strength and could be used as tension piles.

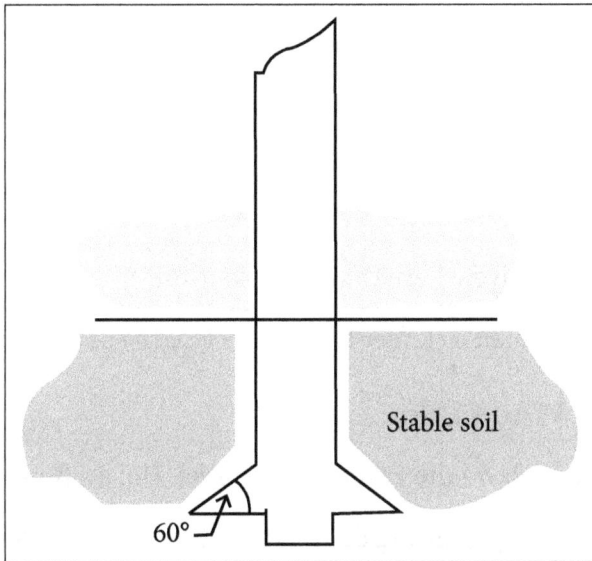

Under-reamed base enlargement to a bore-and-cast-in-situ pile.

Classification of Pile with Respect to Type of Material used

The piles are usually made of timber, concrete or steel. Timber can be used to manufacture temporary piles and when the wood is available at an economical price.

Concrete is used for the manufacturing of pre-cast concrete piles, cast in place and pre-stressed concrete piles, while steel piles are used for permanent or temporary works:

- Timber piles.
- Concrete piles.
- Steel piles.
- Composite piles.

Timber Piles

Used from earliest record time and still used for many works in regions where timber is plentiful. Timber is most suitable for long cohesion piling and piling under the embankments. The timber should be in a good condition and should not been attacked by insects.

For timber piles of length less than 14 meters, the diameter of the tip should be greater than 150 mm. If the length is greater than 18 meters a tip with a diameter of 125 mm is acceptable. It is essential that the timber is driven in the correct direction and should not be driven into firm ground. As this can easily damage the pile.

Keeping it below the ground water level will protect the timber against decay and putrefaction. To strengthen and protect the tip of the pile, timber piles can be provided with toe cover. Pressure creosoting is the usual method of protecting timber piles.

Advantages of Wood Piles

- The piles are easy to handle.

- Relatively inexpensive where timber is plentiful.

- Sections can be joined together and excess length easily removed.

Disadvantages of Wood Piles

- The piles will not above the ground water level. Have a limited bearing capacity.

- Can easily be damaged during driving by stones and boulders.

- The piles are difficult to splice and are attacked by marine borers in salt water.

Concrete Piles

The concrete piles can be divided into pre-cast and cast in place concrete piles.

Prefabricated Concrete Piles or Pre-cast Concrete Piles

It is formed and reinforced in a high-quality controlled concrete, usually used of circle, square, triangle or octagonal section, they are produced in short length in one meter intervals between 3 and 13 meters.

Reinforcement is necessary within the pile to help withstand both handling and driving stresses. Pre stressed concrete piles are also used and are becoming more popular than the ordinary pre cast as less reinforcement is required.

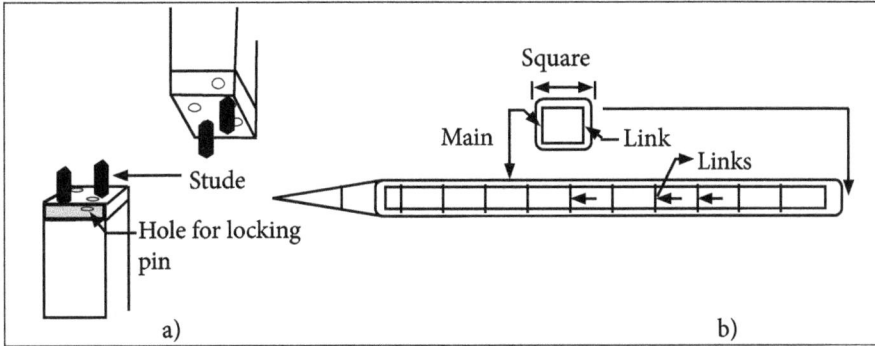

(a) Concrete pile connecting detail (b) Squared pre-cast concert pile.

The Hercules type of pile joint can be easily and accurately casted into a pile and is quickly and safely joined on site. They are made to accurate dimensional tolerances from high grade steels.

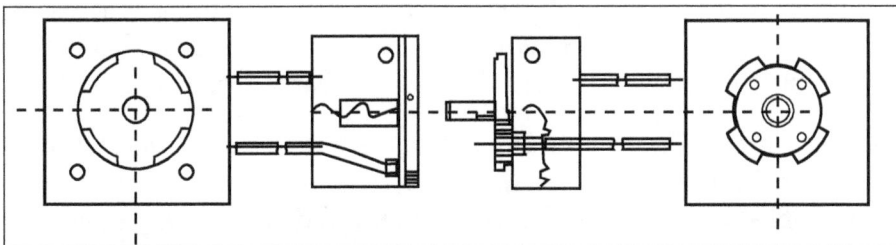

Hercules type of pile joint.

Advantages of Pre - Cast Concrete Piles

- Easy to splice and relatively inexpensive.

- Stable in squeezing ground, for example, soft clays, silts and peats pile material may be inspected before piling.

- Can be driven in long lengths.

- Can increase the relative density of a granular founding stratum.

Disadvantages of Pre- cast Concrete Piles

- Displacement, heave and disturbance of the soil during driving.

- Can be damaged during driving. Replacement piles may be required.

- Cannot be driven with very large diameters or in condition of limited headroom.

Cast in Place Concrete Piles

Cast in place concrete piles are the most commonly used for foundations due to the great diversity available for pouring concrete and the introduction of the pile into the soil. Drilling and driving piles are two types of cast in place concrete piles. However, the implementation of these piles in the place may be accompanied by some problems such as arching, squeezing and segregation.

Piles are poured in tubes with underneath heels and left when lifting the tubes. The types of piles are given below.

Simplex Pile: It is a tube type of cast of diameter 40 cm, has an underneath heel, it is banged underground by an automatic hammer until reaching the arable land for the establishment and then the concrete is poured inside it and is banged by another hammer.

In the meantime, the pipe is raised with a particular amount so as not to enter the soil inside. This pile can bear a maximum of about 40 – 50 tons.

Frankie Pile: It is a number of tubes entering each other in order to easily access to greater depths within the earth. A heel of reinforced concrete can be used and left in the ground to prevent the entry of cold water pipes. This piles can carry a load of 50 – 80 ton.

Vibro Pile: Vibro is a steel pipe of 40 cm in diameter, has a conical heel with a separate flange, it is banged underground by an automatic hammer until reaching the arable land for the establishment and then the heel is removed and put into a tube, after that concrete is poured. The tube is moved up and down in order to compact the concrete. This pile can bear about 60 ton.

Strong Pile: Strong pile is similar to that of the simplex pile except that the bottom heel is made from reinforced concrete covered with cast heel. This pile can carry a load of 25 – 30 ton.

Under Reamed Pile: This pile is used at black clayey soil and lands of non-residual soil, so this soil is highly dangerous to be established on it.

Piles with open tubes without heel, then concrete are poured inside the tube. The pipe has a 40 cm diameter and an average concrete well of 12 to 15 meters depending upon the level of the land to be established. Some these piles are given below.

Strauss Pile: This is very similar to the simplex pile but without a heel. The soil can be removed from the tubes by special devices and concrete is poured instead of the soil. The maximum load that can be carried by these piles ranges from 20 – 25 ton.

Kimbersol Pile: A well is done with a diameter of about 80 cm, until reaching the arable land for the establishment, and then the bottom of the well is compacted using a rounded hammer and then filled with concrete in the ratio of 1: 5, cement and sand respectively. This pile can carry a load of about 80 – 120 ton.

Welfchaulzer Pile: A pipe with a diameter of about 30 – 40 cm is banged until reaching the arable land for the establishment and the inside of soil is removed, then the steel bars are placed and the opened upper hole is covered tightly leaving the holes to connect the compressed air so that the leachates can be expelled, then the concrete is poured by a ratio of 1 : 4.

Raymond Pile: Consists of a cylindrical chip inside each other with a diameter of about 40 – 60 cm at the top of the pile and 20 – 28 cm at the bottom. It is banged inside by a mandrill and the cylindrical chips are leaved in the soil and then filled with concrete.

Advantages of Cast-in-Place Concrete Piles

- Can be inspected before the casting can easily be cut or extended to the desired length.

- Relatively inexpensive.

- The piles can be cast before excavation.

- Pile lengths are readily adjustable.

- An enlarged base can be formed which can increase the relative density of the granular founding stratum leading to much higher end bearing capacity.

- Reinforcement is not determined by the effects of handling or driving stresses.

Disadvantages of Cast-in-Place Concrete Piles

- Heave of neighboring ground surface, which could lead to reconsolidation and the development of negative skin friction forces on piles.

- Tensile damage to unreinforced piles or piles consisting of green concrete, where forces at the toe have been sufficient to resist upward movements.

- Damage piles consisting of uncased or thinly cased green concrete due to the lateral forces set up in the soil. Concrete gets weakened if artesian flow pipes up the shaft of piles when the tube is withdrawn.

- Pre-cast concrete shells or light steel section can be damaged or distorted by hard driving.

- Cannot be driven where headroom is limited.

- Time consuming. It cannot be used immediately after the installation.

- Limited length.

Bored and Cast-in-Place (Non-Displacement Piles)

Advantages

- Length can be readily varied to suit varying ground conditions.

- Can be installed in very large diameters.

- End enlargement up to two or three diameters are possible in clays.

- Material of piles is not dependent on handling or driving conditions.

- Can be installed in very long lengths.

Disadvantages

- Concrete is not placed under ideal conditions and cannot be subsequently inspected.

- Water under artesian pressure may pipe up pile shaft washing out cement.

- Cannot be readily extended above ground level especially in river and marine structures.

- Boring methods may loosen sandy or gravely soils requiring base grouting to achieve the base resistance.

Steel Piles

Made of sectors in the form of H, X or of thick pipes. They are suitable for handling and driving in long lengths. Their relatively small cross-sectional area combined with their high strength makes penetration much easier in firm soil.

They can be easily cut off or joined by welding. If the pile is driven into a soil with a low pH value, then there is a risk of corrosion, but risk of corrosion is not as great as one might think. Although cathode protection or tar coating can be employed in permanent works.

It is common to allow for an amount of corrosion in the design by simply over dimensioning the cross-sectional area of the steel pile. In this way the corrosion process can be prolonged up to 50 years. Normally the speed of corrosion is 0.2-0.5 mm/year and, in design, this value can be taken as 1mm/year.

Advantages of Steel Piles

- The piles are easy to handle and can easily be cut to desire length.

- Can be driven through dense layers. The lateral displacement of the soil during driving is low can be relatively easily spliced or bolted.

- Can be driven hard and in very long lengths.

- Can carry heavy loads.

Disadvantages of Steel Piles

- The piles will corrode.

- Will deviate relatively easy during driving.

- Are relatively expensive.

Composite Piles

Combination of different materials in the same pile. As indicated earlier, part of the timber pile which is installed above the ground water could be vulnerable to insect attack and decay. To avoid this, steel or concrete pile is used above the ground water level, whilst wood pile is installed under the ground water level.

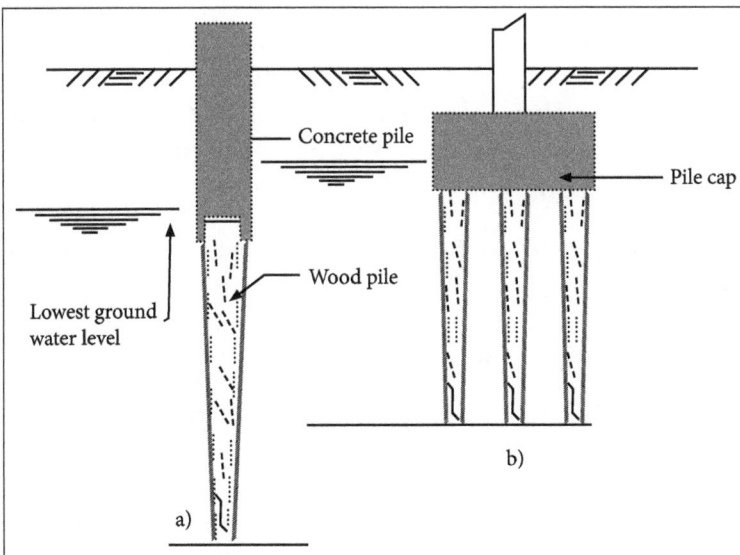

Composite pile.

Protecting Timber Piles from Decay

- By pre-cast concrete upper section above water level.

- By extending the pile cap below water level.

Classification of pile with respect to effect on the soil:

- A simplified division into bored or driven piles is commonly employed.

Driven Piles

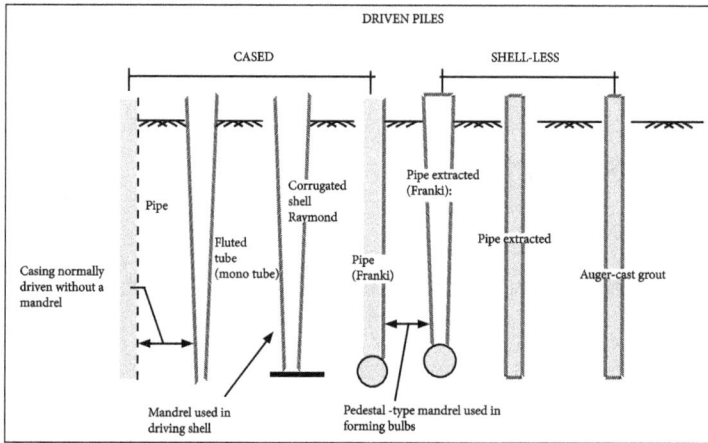

Driven piles.

Driven piles are considered to be displacement piles. In the process of driving the pile into the ground, the soil is moved radially as the pile shaft enters the ground. There can be a component of movement of the soil in the vertical direction.

Bored Piles

Bored piles or replacement piles are normally regarded to be non-displacement piles. A void is formed by boring or excavation before the piles are produced. Piles are produced by casting concrete in the void.

Some soils such as stiff clays are particularly amenable to the formation of piles in this way, because the bore hole walls do not require temporary support except cloth to the ground surface. In unstable ground, such as gravel the ground requires temporary support from bentonite slurry or casing.

Alternatively the casing may be permanent, but driven into a hole which is bored as casing is advanced.

The different technique, that is essentially non-displacement, is to intrude a grout or a concrete from an auger which is rotated into the granular soil and hence produce a grouted column of soil.

There are three non-displacement methods: grout or concrete intruded piles, bored cast-in-place piles, particularly pre-formed piles.

The following are replacement piles:

- Augered.
- Cable percussion drilling.
- Large-diameter under-reamed.
- Types incorporating pre caste concrete unite.
- Drilled-in tubes.
- Mini piles.

4.2   Factors Influencing the Selection of Pile

Piles are constructed in a group of vertical, batter or a combination of vertical and batter piles. The distribution of loads applied to a pile group are transferred non-linearly and indeterminately to the soil. Interaction effects between adjacent piles in a group lead to complex solutions. Factors that are considered, affect the resistance of the pile group to movements and also the transfer of load through the pile group to the soil.

1. Soil Modulus

- The elastic soil modulus Es and the laterals modulus of subgrade reaction Els relate lateral, axial and rotational resistance of the pile-soil medium to displacements. Seepage pressures and water table depth affect the modulus of cohesion less soil.

- The modulus of submerged sands should be reduced by the ratio of the submerged unit weight divided by the soil unit weight.

2. Batter

- Battered piles are used in the groups at least two or more piles to increase capacity and loading resistance. The angle of inclination must rarely exceed 20 degrees from the vertical for normal construction and should never exceed $26\frac{1}{2}$ degrees.

- These piles should be avoided where significant negative skin friction and down drag forces may occur. This pile should be avoided where the structure's foundation must respond with ductility to unusually large loads or where large seismic loads could be transferred to the structure through the foundation.

3. Fixity

- The fixity of the pile head into the pile cap can influence the loading capacity of the pile group. This pile rather than pinning into the pile cap generally increases the lateral stiffness of the group and the moment.

- Therefore, a group of fixed piles can support about twice the lateral load at identical deflections as the pinned group. A fixed connection between the pile and the cap is also able to transfer the significant bending moment through the connection.

- The minimum vertical embedment distance of the top of the pile into the cap required for achieving a fixed connection is 2B. Where, B is the pile width or diameter.

4. Stiffness of Pile Cap

- The stiffness of the pile cap will influence the distribution of the structural loads to the individual piles.

- The thickness of the pile cap must be at least four times the width of the individual pile which causes a significant influence on the stiffness of the foundation.

- A rigid cap may be assumed if the stiffness of the cap is 10 or more times greater than the stiffness of the individual piles, as basically true for massive concrete caps. A rigid cap can generally be assumed for gravity type hydraulic structures.

5. Nature of Loading

- Cyclic, static, dynamic and transient loads affect the ability of the pile group to resist the applied forces. Cyclic, vibratory or repeated static loads can cause greater displacements than a sustained static load of the same magnitude. In some cases displacements can double.

6. Driving

- The apparent stiffness of a pile in a group can be because of the density of the soil within and around a pile group may be increased by driving.

- The pile group as a whole will not reflect this increased stiffness, since the soil around and outside the group may not be favorably affected by driving and displacements larger than that anticipated may occur.

7. Sheet Pile Cutoffs

- Sheet pile cutoffs enclosing a pile group load capacity. The length of the cutoff can be determined from a flow net or other seepage analysis.

- The net pressure acting on the cutoff is the sum of the unbalanced earth and water pressures caused by greater than that of an isolated pile driven into cohesion less soil may change the stress distribution in the soil and influence the cutoff.

- Steel pile cutoffs must be considered in the analysis as not totally impervious. Flexible steel sheet piles will cause negligible load to be transferred to the soil.

- The rigid cutoffs, such as a concrete cutoff, will transfer the unbalanced earth pressure and water pressure to the structure and can be accounted for in the analysis of the pile group.

8. Interaction Effects

- Deep foundations where the spacing between the individual piles is less than six times the pile width B which cause interaction effects between the adjacent piles from overlapping of stress zones in the soil, figure in-situ soil stresses from pile loads are applied over a much larger area and extend to a greater depth leading to a greater settlement.

9. Pile Spacing

- The piles in a group should be spaced such that the bearing capacity of the group is optimum. The optimum spacing for driven piles is about 3 to 3.5B or 0.02L + 2.5B, where L is the embedded length of the piles. Pile spacings should be at least 2.5 B.

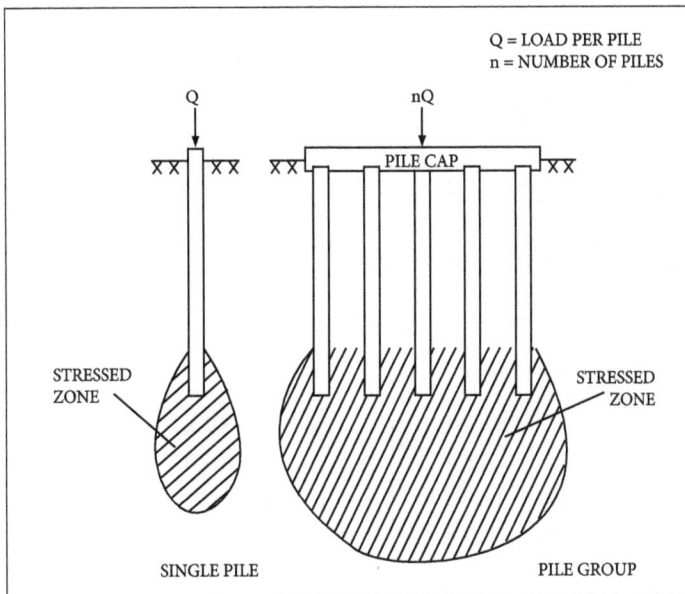

Stress zones in soil supporting piles.

4.3   Carrying Capacity of Single Pile in Granular and Cohesive Soil

Load Carrying Capacity of Piles

The ultimate bearing capacity or the ultimate bearing resistance or ultimate load carrying capacity Q_{up} of the pile is defined as the maximum load which can be carried by a pile and at which the pile continues to sink without further increase of the load.

The allowable load Q_a is the safe load which the pile can carry safely and is determined on the basis of: ultimate bearing resistance divided by suitable factor of safety, (ii) the permissible settlement and (iii) overall stability of the pile-foundation. The load carrying capacity of a pile can be evaluated by the following methods:

- Pile load tests.

- Dynamic formulae.

- Penetration tests.

- Static formulae.

There are various methods of evaluating the load carrying capacity of the single piles. They are:

- Based on CPT Results.

- By Pile Loading Tests.

- Based on SPT Results.

- By Pile Driving Formula.

- By Static Bearing Capacity Equations.

Static Bearing Capacity Equations

The static method gives the ultimate bearing capacity of an individual pile, based on the characteristics of soil.

Assume a pile of diameter D is driven to a length L from the ground surface. Let Q_u be the net load applied to the pile head. The weight of the pile is W and unit weight of soil is Y. If q_b is the unit base friction and q_s is the unit skin friction respectively, then we can write the ultimate capacity of pile as,

$$Q_u + W - \gamma D_f = q_b A_b + q_s A_s \quad ...(1)$$

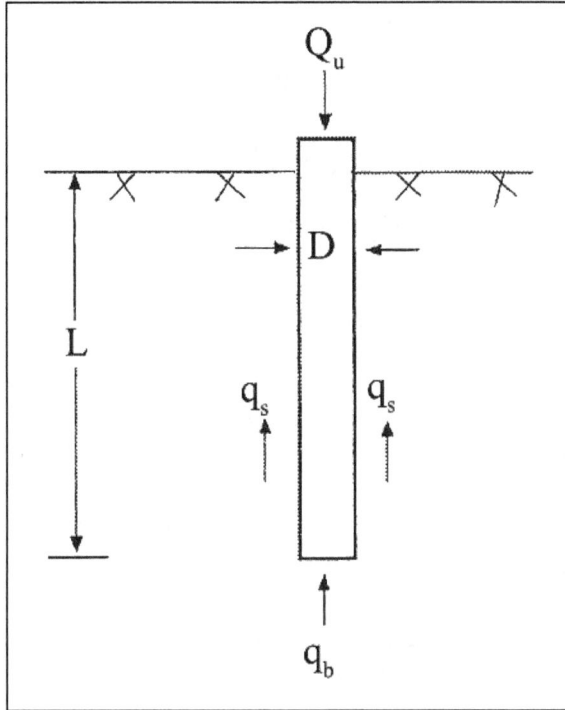

Vertical Loads on Single Pile.

If the weight of the pile and weight of the displaced soil are assumed to be equal we have,

$$Q_u = q_b A_b + q_s A_s$$

Then,

$$q_b = \tfrac{1}{2}\, B\gamma N\gamma + \sigma_v N_q + s_u N_c - \sigma_v$$

or,

$$q_b = \tfrac{1}{2}\, B\gamma N\gamma + \sigma_v\left(N_q - 1\right) + s_u N_c \qquad \ldots (2)$$

Similarly unit skin friction can be written as,

$$q_s = s_a + K\sigma_v \tan \delta \qquad \ldots (3)$$

Where, σv is the effective normal stress acting along the shaft of the pile and the variation of the normal stress is linear. sa is the adhesion between the pile and the surrounding soil. The average effective normal stress is $(0 + \sigma_v)/2 = \sigma_v/2 = \sigma'_v$.

Hence,

$$q_s = s_a + K \sigma'_v \tan\delta \qquad \ldots (4)$$

Therefore, the net ultimate pile capacity becomes,

$$Q_u = \left[\frac{1}{2} B\gamma N\gamma + \sigma_v \left(N_q - 1 \right) + s_u N_c \right] Ab + \left[s_a + K\sigma'_v \tan\delta \right] A_s \qquad ...(5)$$

This is the general equation applicable to soil having both cohesion and friction.

Cohesion less soil – If the soil is cohesion less (i.e.) cohesion is zero then the equation (5) reduces to,

$$Q_u = \sigma_v \left(N_q - 1 \right) Ab + K\sigma'_v \tan\delta A_s \qquad ...(6)$$

The equation (6) neglects the term $\frac{1}{2} B\gamma N\gamma$, since pile width is very small as compared to its length. The value of N_q is obtained by the relationship suggested by Berezantsev (1961), which is shown in figure. To use the graph shown in figure. The value of ϕ should be known and this can be obtained from the results of Standard Penetration Test using the graph given in figure. The suggested relation can only be used when the piles are driven to depths that are greater than five times the pile width if less use Terzaghi's bearing capacity equation. The term N_q is very sensitive to ϕ a relatively small error in estimating ϕ will lead to much larger error in N_q.

Meyerhof has suggested that the bearing capacity of the piles driven into loose sands can be doubled owing to compaction. The suggested value of K with δ is given in table below:

Table: Values of K and δ.

Pile material	Δ	K	
		Loose	Dense
Steel	20	0.5	1
Concrete	0.75 ϕ	1	2
Wood	0.67 ϕ	1.5	3

Cohesive soil– In cohesive soil, the angle of internal friction is zero. Hence, the above equation (5) can be written as,

$$Q_u = s_u N_c A_b + s_a A_s \qquad ...(7)$$

In the above equation s_a is the adhesion between the surrounding soil and the pile. S_u is the un drained shear strength at the base. Driving piles in clay decreases the shear strength in many ways. If the reduction factor is α, then we can write,

$$\alpha = s_a / s'_u \qquad ...(8)$$

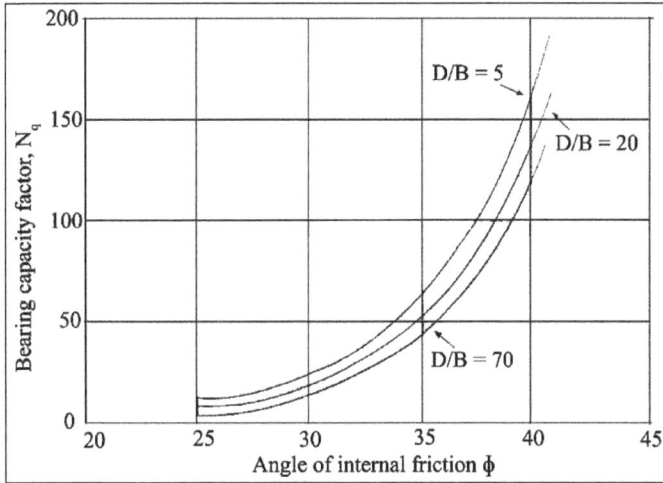

Berezantsev's Bearing Capicity Factor Nq.

Where, s'_u = Average undrained strength over the embedded length of pile,

$$s_a = \alpha s'_u \qquad \qquad ...(9)$$

Hence from equations (7) and (9),

$$Q_u = s_u N_c A_b + \alpha s'_u A_s \qquad \qquad ...(10)$$

In the above equation, the value of N_c is taken equal to 9. In clayey soil base resistance can be neglected. So an error in the estimation of base load has less significance. Therefore, the main objective is the determination of adhesion factor mobilized between the pile and the surrounding clay.

4.4   Static Formula

Load Carrying Capacity of Pile using Static Analysis

There are two ways in which the pile transfers the load to the soil. Firstly, through the tip-in compression, known as "point-bearing" or "end-bearing". Secondly, by shear along the surface termed as "skin friction".

Load Carrying Capacity of the Cast in-Situ Piles in the Cohesive Soil

The ultimate load carrying capacity (Q_u) of pile in cohesive soils is given by the below formula, where, the first term represents the end bearing resistance (Q_b) and the second term gives the skin friction resistance (Q_s),

$$Q_u = \left[A_p N_c c_p \right] + \left[\sum_{i=1}^{n} \alpha_i \, c_i \, A_{si} \right]$$

Where,

A_p = Cross-sectional area of pile tip, in m².

Q_u = Ultimate load capacity, kN.

N_c = Bearing capacity factor, may be taken as 9.

a_i = Adhesion factor for the ith layer depending on the consistency of the soil. It depends on the undrained shear strength of soil and may be obtained from the below figure.

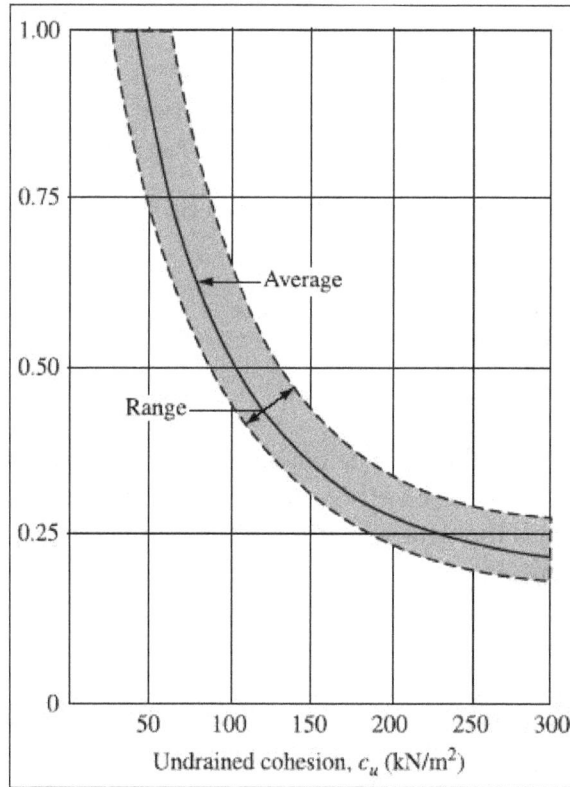

Variation of alpha with cohesion.

c_i = Average cohesion for the ith layer, in kN/m².

A_{si} = Surface area of pile shaft in the ith layer, in m².

A minimum factor of safety of 2.5 is used to arrive at the safe pile load capacity (Q_{safe}) from ultimate load capacity (Q_u),

$$Q_{safe} = Q_u/2.5$$

Load Carrying Capacity of Cast in-situ Piles in Cohesion Less Soil

The ultimate load carrying capacity of pile, "Q_u", comprises of two parts. One part is due to friction, known as shaft friction or skin friction or side shear denoted as "Q_s" and the other is due to end bearing at the tip or base of the pile toe, "Q_b".

The equation given below can be used to calculate the ultimate load carrying capacity of the pile,

$$Q_u = A_p \left(\frac{1}{2} D\gamma N\gamma + P_D N_q \right) + \sum_{i=1}^{n} k_i P_{Di} \tan \delta A_{si}$$

Where,

D = Diameter of pile shaft, m.

A_p = Cross-sectional area of the pile base, m².

γ = Effective unit weight of the soil at pile tip, kN/m³.

N_γ= Bearing capacity factor.

N_q = Bearing capacity factor.

φ = Angle of internal friction at pile tip.

PD = Effective overburden pressure at pile tip, in kN/m².

K_i = Coefficient of earth pressure applicable for the ith layer.

δ_i = Angle of wall friction between pile and soil for the ith layer.

PD_i = Effective overburden pressure for the ith layer, in kN/m².

As_i = Surface area of the pile shaft in the ith layer, in m².

The first term is the expression representing the end bearing capacity of the pile (Q_b) and the second term is the expression to represent skin friction capacity of the pile (Q_s).

A minimum factor of safety of 2.5 can be used to arrive at the safe pile capacity (Q_{safe}) from ultimate load capacity (Q_u),

$Q_{safe} = Q_u/2.5$

The value of bearing capacity factor N_q is obtained from the figure.

For driven piles in loose to dense sand with φ varying between 300to 400, k_i values in the range of 1 to 1.5 may be used.

δ, the angle of wall friction can be taken equal to the friction angle of the soil around the pile stem.

The maximum effective overburden at the pile base must correspond to the critical depth, which may be taken as 15 times the diameter of the pile shaft for $\varphi \leq 300$ and increasing to 20 times for $\varphi \geq 400$

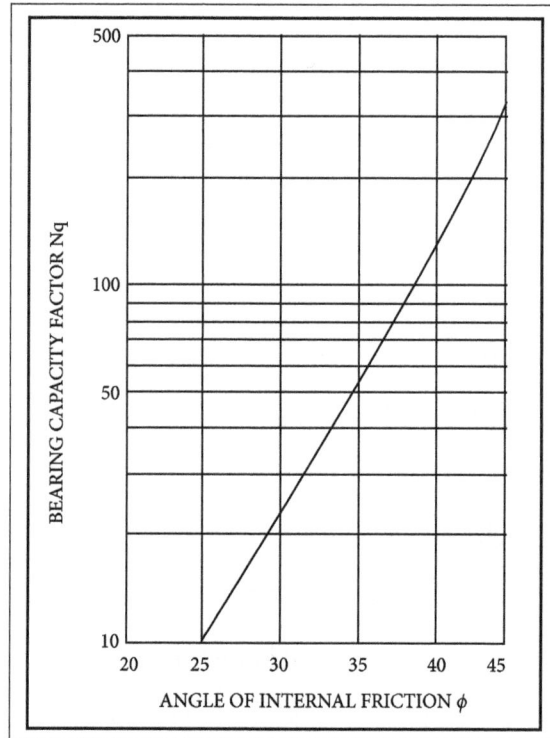

Bearing factor value.

The value of the bearing capacity factor N_γ is computed using the equation given below,

$$N_\gamma = 2\left(N_q + 1\right) \tan\phi$$

For piles passing through cohesive strata and terminating in a granular stratum, a penetration of at least twice the diameter of the pile shaft should be given into the granular stratum.

Problems

1. A wooden pile is being driven with a drop hammer weighing 20 kN and having a free fall of 1.0 m. The penetration in the last blow is 5 mm. Let us determine the load carrying capacity of the pile according to the Engineering News Formula.

Solution:

Given:

$W = 20 \text{ kN}$

$H = 1 \times 100 = 100 \text{cm}$

$$S = 0.5 \text{ cm}$$

$$C = 2.5 \text{ cm}$$

To find: Load carrying capacity.

Formula to be used,

$$Q_a = \frac{WH}{6(S+C)}$$

Where W = 20 kN ; H= 1 × 100 = 100 cm; S = 0.5 cm; C = 2.5 cm,

$$Q_a = \frac{20 \times 100}{6(0.5+2.5)} = 111.1 \text{ kN}$$

2. In a 16 pile group, the pile diameter is 45 cm and center to center spacing of the square group is 1.5 m. If c = 50 kN/m², here let us determine whether the failure would occur with the pile acting individually or as a group. Also neglect bearing at the tip of the pile. All piles are 10 m long. Take m = 0.7 for shear mobilization around each pile.

Solution:

Given:

$$n = 16$$

$$d = 45 \text{ cm}$$

$$L = 10 \text{ m}$$

$$\text{Spacing} = 1.5 \text{ m}$$

$$c = 50 \text{ kN}/\text{m}^2$$

Formula to be used,

$$Q_{ug} = c \times \text{perimeter} \times \text{length}$$

$$Q_{ug} = n \, Q_{up} = n \left\{ mcA_p \right\}$$

$$A_p = \pi \, dL$$

Width of group $= B = (150 \times 3) + 45 = 495 \text{ cm} = 4.95 \text{ m}$

For the Group:

$$Q_{ug} = c \times \text{perimeter} \times \text{length} = c \times 45 \times L = 50 \times 4 \times 4.95 \times 10 = 9900 \text{ kN}$$

For the Piles Acting Individually:

$$Q_{ug} = n\,Q_{up} = n\,\{mcA_p\}$$

$$A_p = \pi\,dL = \pi \times 0.45 \times 10$$

$$Q_{ug} = 16 \times 0.7 \times 50 \times \pi \times 0.45 \times 10 = 7917\ kN$$

Which is less than the load carried by the group action. Hence the foundation will fail by the piles acting individually and the load at failure would be 7917 kN.

4.5   Dynamic Formulae

When a pile hammer hits the pile, the total driving energy is equal to the weight of hammer times the height of drop or stroke. In addition to this, in the case of double acting hammers, some energy is also imparted by the steam pressure during the return stroke.

The total downward energy is consumed by the work done in penetrating the pile and by certain losses. The various dynamic formulae are essentially based on this assumption.

It is also assumed that soil resistance of dynamic penetration of pile is the same as to the penetration of pile under static or sustained loading. Following are some of the commonly used dynamic formulae.

Engineering News Formula

The Engineering News formula was proposed by A.M. Wellington (1818) in the following general form,

$$Q_a = \frac{WH}{F(S+C)}$$

Where,

Q_a = Allowable load

W = Weight of hammer

H = Height of fall

F = Factor of safety = 6

S = Final set per blow, usually taken as average penetration, cm per blow for the last 5 blows of a drop hammer or 20 blows of a steam hammer.

C = Empirical constant

Denoting W in kg, H in cm, 5 in cm

C = 2.5 cm for drop hammer= 0.25 cm for single and double acting hammers.

The above formula reduces to the following forms:

Drop Hammers:

$$Q_a = \frac{WH}{6(S+2.5)}$$

Single acting steam hammers:

$$Q_a = \frac{WH}{6(S+0.25)}$$

Double acting steam hammers:

$$Q_a = \frac{(W+ap)H}{6(S+0.25)}$$

Where,

a = Effective area of piston (cm²).

p = Mean effective steam pressure (kg/cm²).

Hiley 's Formula

Indian Standard IS: 2911 (Part I) 1964 gives the following formula based of original expression of Hiley,

$$Q_f = \frac{\eta_h \ WH \ \eta_b}{S + \dfrac{C}{2}}$$

Where,

Q_f = Ultimate load on pile.

W = Weight of hammer in kg.

H = Height of drop of hammer in cm.

S = Penetration or set, in cm per blow.

C = Total elastic compression C1 + C_2 + C_3.

C_1, C_2, C_3 = Temporary elastic compression of dolly and packing, pile and soil respectively.

η_h = Efficiency of hammer, variable from 65 per cent for some double acting steam hammers to 100 percent for drop hammers released by trigger.

η_b = Efficiency of hammer blow (i.e. ratio of the energy after impact to striking energy of ram),

$$\eta_b = \frac{W + e^2 P}{W + P} \left(\text{For the case of } W > eP\right)$$

$$\eta_b = \frac{W + e^2 P}{W + P} - \left\{\frac{W - eP}{W + P}\right\}^2 \left(\text{for the case when } W < eP\right)$$

Where,

P = Weight of pile

e = Coefficient of restitution (variable from zero for a timber pile with poor condition of head or for excess packing in the driving cap to 0.5 for double acting hammer driven steel piles without driving cap or reinforced concrete piles without helmet but with packing on top.)

These equations are applicable for friction piles. For piles driven to refusal on rock (end bearing pile) a value of 0.5 P is substituted in above expressions.

The product H is sometimes referred to as the effective fall of the hammer.

For double acting hammers the rated energy in the same length unit as S and C is substituted for W H. The allowable load is obtained by using a factor of safety 2 or 2.5.

Comments about the use of Dynamic Formulae

- Dynamic formulae are best suited to coarse grained soils for which the shear Strength is independent of rate of loading, because they allow no development of excess pore pressure around the pile during driving if saturated or dry.

- The great objection to any of the pile driving formulae is the uncertainty about the relationship between the dynamic and static resistance of soil.

- In case of submerged loose uniform fine sands, impact of driving may cause liquefaction of soil, thus showing much less resistance than that which will occur under a static load. Similarly, a very dense saturated fine sand may show an increased driving resistance which decreases.

- For clays, the dynamic formulae are valueless because the skin friction developed in clay during driving is very much less than which occurs after a period of time.

Also, the point resistance is much more at the time of driving because of pore pressure developed in clay, which reduces later on when the pore pressure dissipate:

- Dynamic formulae give no indication about probable fixture settlement or temporary changes in soil structure.

- The formulae do not take into account the reduced bearing capacity of pile when in a group.

- Law of impact used for determining energy loss is not strictly valid for piles subjected to restraining influence of the surrounding soil.

- In Engineering News formula, the weight of the pile and hence its inertia effect is neglected.

- Energy losses due to vibrations, heat and damage to dolly or packing are not accounted for.

- In Hiley's formula, a number of constants are involved, which are difficult to determine.

4.6   Capacity from In-Situ Tests

Cohesion Less Soil

In cohesion less soils, the bearing capacity is extremely high with respect to shear failure criteria. In sands, the shear failure criteria govern the capacity only in the case of very narrow footing located in loose sand below the water table. In most of the cases the bearing capacity in sands is governed by settlement criteria.

Shear Failure Criteria

Teng suggested the bearing capacity in sands for a strip, circular or square and raft foundation from the results of the standard penetration tests. The proposed equations are as follows.

Strip Footing

The bearing capacity of the strip footing is given as,

$$q_{na} = 0.0167 \, N^2 BR_{W_1} + 0.0277 \left(100 + N^2\right) Df \, R_{W_2} \quad \ldots (1)$$

Square or circular footing – The bearing capacity of circular or square footing is given as,

$$q_{na} = 0.011 \, N^2 BR_{W_1} + 0.033 \left(100 + N^2\right) D_f \, R_{W_2} \quad \ldots (2)$$

Raft Foundation

The bearing capacity of the raft foundation is given as,

$$q_{na} = 0.02\,N^2BR_{W_1} + 0.06\left(100 + N^2\right)D_f\,R_{W_2} \qquad ...(3)$$

Where,

   N = Corrected SPT value.

   D_f = Depth of foundation.

   B = Width of foundation.

   R_{W_1}, R_{W_2} = Water table correction factors.

The unit of q_{na} is t/m^2.

Settlement Criteria

In sand, in most of the cases the bearing capacity is governed by the settlement criteria. The design philosophy for footings on sands is explained below.

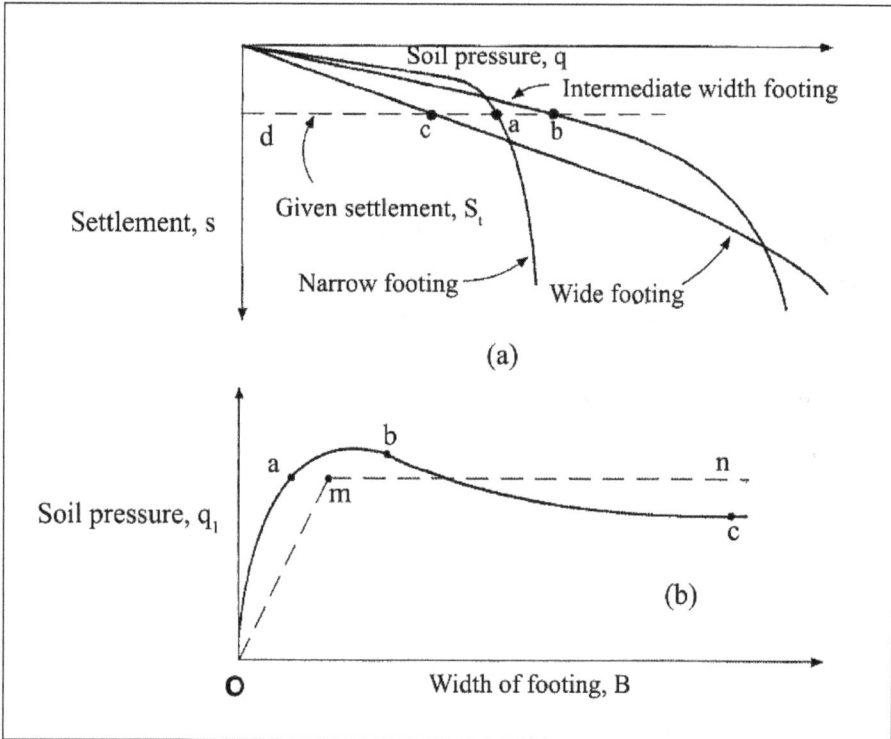

Footings on sands.

Typical load settlement relationships for the footings of increasing sizes resting on a homogeneous deposit of sand of a certain relative density.

The load settlement curves show that the ultimate bearing capacity increases as the width of the foundation increases. However, for a given permissible settlement, say S, the pressure required to cause it greater for footing having intermediate width than for the footing having larger width.

The pressure corresponding to the three widths such as narrow, intermediate and large are represented in the figure by the points a, b and c respectively. The curves also show that, in case of a small footing, a small increase in load can cause the footing to settle excessively and subsequently leads to bearing capacity failure.

While in the ease of larger footings, a small change in the load does not have significant influence on the settlement value.

Footings on the granular soils are generally proportioned by the use of N values. Most of the methods propose empirical equations or charts to determine the safe bearing pressure for a specified maximum total settlement in terms of N values.

4.7   Negative Skin Friction

Negative skin friction is a down ward drag acting on a pile due to the downward movement of the surrounding compressible soil relative to the pile. This happens when the surrounding compressible soil has been recently filled or formed.

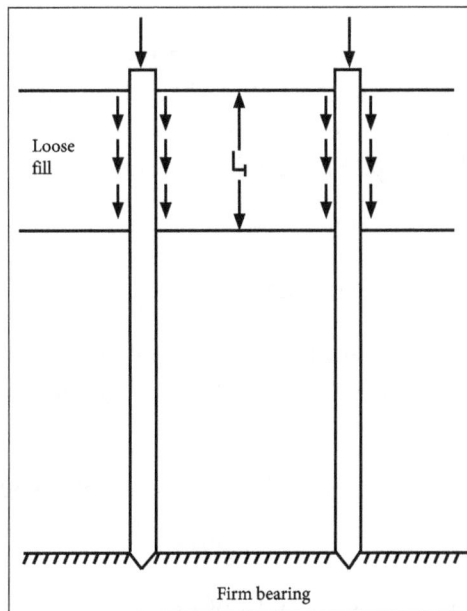

Negative skin friction.

As the soil consolidates, the earth fill moves downward, developing friction forces on the perimeter of the pile which tend to carry the pile farther to the ground.

The negative skin friction can also be developed by the lowering of the ground water, the increase in effective stress causing consolidation of the soil, with the resultant settlement and friction force being developed on the pile.

For individual piles, the magnitude of negative friction Q_{nf} may be taken as follows:

For Cohesive Soils:

$$Q_{nf} = p \cdot c \cdot L_f$$

For Granular Soils:

Where,

  p = Perimeter of the pile.

  L_f = Depth of fill or soil which is moving vertically.

  K = Earth pressure coefficient ($K_a < K < K_p$).

  y = Unit weight of soil.

  c = Cohesion of soil in zone of L_f.

  f = Coefficient of friction $\approx \tan \varphi$.

Problems

1. A n-pile group has to be proportioned in a uniform pattern in soft clay with equal spacing in all directions. Assuming any value of c, let us determine the optimum value of spacing of piles in the group. Take n = 25 and m = 0.7 and also neglect the end bearing effect and assume that each pile is circular in section.

Solution:

To find: Optimum value of spacing of piles.

The optimum spacing of individual piles is based on the premise that the efficiency of the pile group is unity. That is the total load carried by the piles, acting individually is equal to the load carried by group action.

Let the center to center spacing of each piles = Diameter of each pile = d,

  n = 25

$\therefore$ Width for each pile group 4 s + d; Length of pile = L

$\therefore$ Load carried by group action is,

$$Q_{ug_1} = c \left\{ 4(4s + d) L \right\} = 4Cl \left(4s + d \right) \qquad ...(i)$$

Load carried by pile acting individually is,

$$Q_{ug2} = n(mc)(\pi dL) = 25 \times 0.7\, c\pi dL = 55dL \cdot c \qquad ...(ii)$$

Equating (i) and (ii) for the optimum spacing,

$$4cL(4s+d) = 55dL \cdot c$$

or,

$$16s + 4d = 55d$$

$$\therefore s = \frac{51}{16}d = 3.19\ d.$$

Hence the spacing should be equal to 3.19 times diameter of the pile.

2. A reinforced concrete pile weighing 30 kN (inclusive of helmet and dolly) is driven by a drop hammer weighing 40 kN and having an effective fall of 0.8 m. The average set per blow is 1.4 cm. The total temporary elastic compression is 1.8 cm. assuming the co-efficient of restitution as 0.25 and a factor of safety of 2, let us determine the ultimate bearing capacity and the allowable load for the pile.

Solution:

Given:

$$P = 30\ kN;\ W = 40\ kN;\ \eta_h\ H = 0.8\ m = 80\ cm;\ S = 1.4\ cm; C = 1.8\ cm\ e = 0.25;\ F = 2$$

To find:

- Ultimate bearing capacity.
- Allowable load for the pile.

Formula to be used,

$$\eta_b = \frac{W + e^2\,P}{W + P}$$

$$Q_f = \frac{(\eta_h\,H)\,W\eta_b}{S + \dfrac{C}{2}}$$

$$Q_a = \frac{Q_f}{F}$$

Since,

$$W > eP,$$

$$\eta_b = \frac{W + e^2 P}{W + P} = \frac{40 + 30(0.25)^2}{40 + 30} = 0.597$$

$$\therefore Q_f = \frac{(\eta_h H) W \eta_b}{S + \dfrac{C}{2}} = \frac{80 \times 30(0.25)^2}{1.4 + \dfrac{1.8}{2}} = 830 \text{ kN}$$

$$\therefore Q_a = \frac{Q_f}{F} = \frac{830}{2} = 415 \text{ kN}.$$

3. A friction pile group to carry a load of 3000 kN including the weight of the pile cap at a site where the soil is a uniform clay to a depth of 20 m, underlain by rock. Average unconfined compressive strength of the clay is 70 kN/m². The clay can be assumed to be of normal sensitivity and normally loaded with liquid limit 60%. A factor of safety of 3 is required against shear failure.

Solution:

Given:

Load = 3000 kN

Depth = 20 m

Strength of the clay = 70 kN/m²

Liquid limit 60%

Formula to be used,

$$c = \frac{q_u}{2}$$

$$Q_{ug} = n \, c \, \pi \, dL$$

$$B = 3s + d$$

$$A_b = B \times B = B^2$$

$$c = \frac{q_u}{2} = \frac{70}{2} = 35 \text{ kN/m}^2;$$

Permissible $c = \dfrac{35}{3} \text{ kN/m}^2.$

Let the length of pile = 10 m; Diameter of pile = 0.5 m.

Spacing of pile = 3 d = 150 cm (say). Let the number of piles = n Considering the piles to act individually, the load at failure is given by,

$$Q_{ug} = n \, c \, \pi \, dL$$

$$\therefore 3000 = n \times \frac{35}{3} \times \pi \times 0.5 \times 10$$

$$\therefore n = \frac{3 \times 3000}{35 \times \pi \times 0.5 \times 10} = 16.37$$

For a square arrangement, keep n = 16.

The modified length L will then have to be increased by the ratio $\frac{16.37}{16}$,

$$\therefore L = 10 \times \frac{16.37}{16} = 10.23 \text{ m.}$$

Adopt L = 11 m.

Check for the group action,

$$B = 3s + d = 3 \times 150 + 50 = 500 \text{ cm} = 5 \text{ m}$$

∴ Load taken by group action =4 BL. c + Ap cNc. (Taking into account the end bearing),

$$A_b = B \times B = B^2 = (5)^2 = 25 \text{ m}^2; \ N_c = 9$$

$$\therefore Q_{ug} = \left(4 \times 5 \times 11 \times \frac{35}{3}\right) + \left(25 \times \frac{35}{3} \times 9\right) = 2566.7 + 2625 = 5191.7 \text{ kN}$$

This is greater than 3000 kN.

Hence safe.

4.8 Uplift Capacity

The uplift capacity of a pile group, when the vertical piles are arranged in a closely spaced groups may not be equal to the sum of the uplift resistances of the individual piles. This is because, at ultimate load conditions, the block of soil enclosed by the pile group gets lifted.

The manner in which the load is transferred from the pile to the soil is quite complex. A simple way of calculating the uplift capacity of a pile group embedded in cohesion less soil is a spread of load of 1 Horizontal: 4 Vertical from the pile group base to the ground surface may be taken as the volume of the soil to be lifted by the pile group.

For simplicity in calculation, the weight of the pile embedded in the ground is assumed to be equal to the volume of soil it displaces. If the pile group is partly or fully submerged, the submerged weight of the soil below the water table has to be taken.

In the case of cohesive soil, the uplift resistance of the block of soil in the untrained shear enclosed by the pile group given in has to be considered. The equation for the total uplift capacity P_{gu} of the group may be expressed by,

$$P_{gu} = 2L\left(\overline{L} + \overline{B}\right)\overline{c}_u + W$$

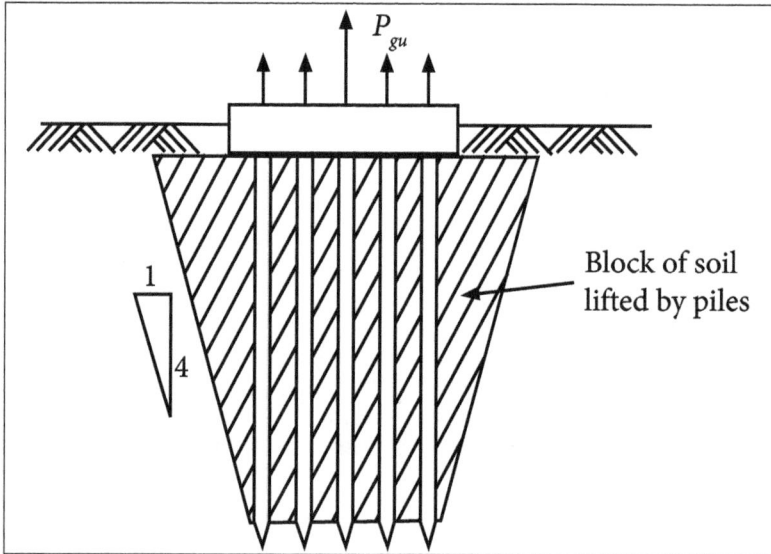

(a) Uplift of a group of closely-spaced piles in cohesionless soil.

Uplift capacity of a pile group.

A factor of safety of 2 may be used in both cases of piles in clay as well as in sand.

The uplift efficiency E_{gu} of a group of piles may be expressed as,

$$E_{gu} = \frac{P_{gu}}{nP_{us}}$$

Where,

n = Number of piles in the group.

P_{us} = Uplift capacity of a single pile.

The efficiency E_{gu} varies with the method of installation of the piles, spacing, length and the type of soil.

The available data indicates that E_{gu} increases with the spacing of the piles. Meyerhof and Adams presented some data on uplift efficiency of groups of two and four model circular footings in the clay.

The results indicate that the uplift efficiency increases with the spacing of the bases or footings and as the depth of embedment decreases, but decreases as the number of footings or bases in the group increases.

How far the footings would represent the piles is a debatable point. For uplift loading on pile groups in sand, there appears to be a little data from full scale field tests.

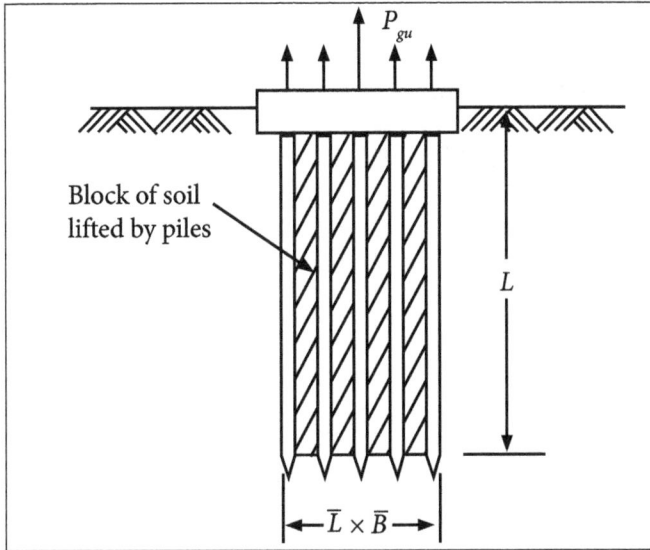

(b) Uplift of a group of piles in cohesive soils.

4.9 Group Capacity by Different Methods

Efficiency of Pile Group

When several closely spaced piles are grouped together, it is reasonable to expect that the soil pressure developed in the soil as resistance will overlap. The bearing capacity of a pile group may or may not be equal to the sum of the bearing capacity of individual piles constituting a group.

Theory and tests have shown that the total bearing value Q_{ug} of a group of friction piles, particularly in clay, may be less then the product of the friction bearing value of an individual pile multiplied by the number of piles n in a group.

However, there is no reduction due to grouping occurs in end bearing piles. For combined end bearing and friction piles, only the load carrying capacity of the frictional portion is reduced. A method of estimating the bearing capacity of a group of friction piles is to multiply the quantity n Q_{up} by a reduction factor called the efficiency of pile group,

$$Q_{ug} = n\, Q_{up}.\eta_g$$

Where,

Q_{ug} = Load carried by group of friction piles.

Q_{uP} = Load carried by each friction pile.

n = Number of piles.

η_g = Efficiency of pile group.

The efficiency of pile group depends upon the following factors: Characteristics of pile such as length, diameter, material etc., spacing of pile, total number of piles in a row and number of rows etc. A number of formulae are available for determining the efficiency of pile group. Some of these are given below:

Converse Labarre Formulae

$$\eta_g = 1 - \frac{\theta}{90}\left[\frac{(n-1)m+(m-1)n}{mn}\right]$$

Where,

m = Number of rows.

n = Number of piles in a row.

θ = $\tan^{-1}$ (d/s) (degrees).

d = Diameter of pile.

s = Spacing of pile.

Seiler-Keeney formulae,

$$\eta_g = \left[1-0.479\left(\frac{s}{s^2-0.093}\right)\left(\frac{m+n-2}{m+n-1}\right)\right]+\frac{0.3}{m+n}$$

Where, s = Average spacing, center to center, in meters.

Feld's-Rule

According to this rule, the value of each pile is reduced by one-sixteenth on account of the effect of the nearest pile in each diagonal or straight row of which the pile in question is a member.

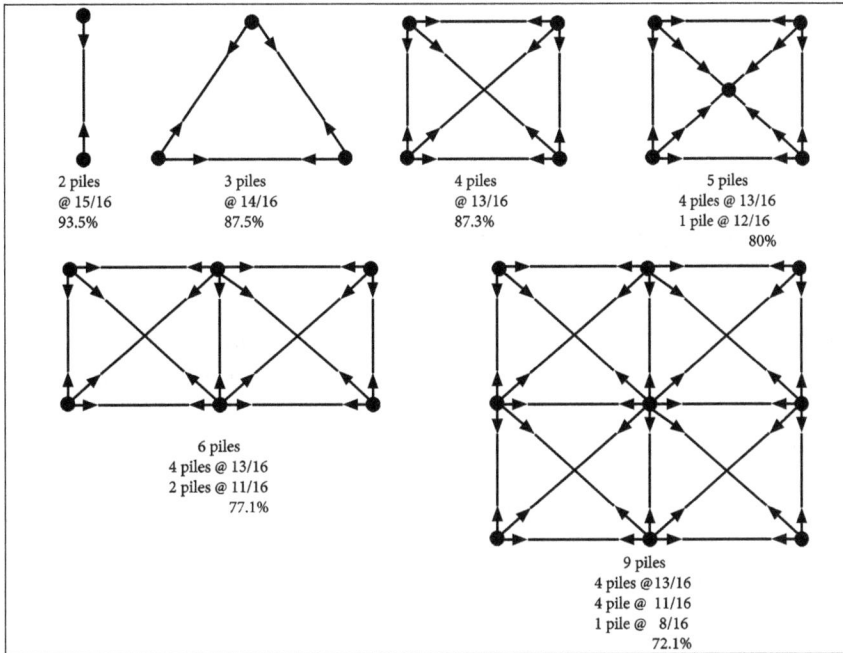

Efficiency of pile groups by feld's rule.

Design of Pile Groups

The bearing capacity of single pile in clay is mainly due to friction and the point bearing resistance may be negligible. In a pile group, the piles are connected at its top by a pile cap which is rigid.

Hence the failure of a pile group is likely to occur at a load which may be smaller than the ultimate load carried by each pile multiplied by the number of piles in the group. The area of the pile group, along failure surface, is approximately equal to the perimeter P of the pile group multiplied by the length L of the pile.

The ultimate load will then be equal to,

$$Q_{ug} = PL_{rf} + A._{rp}$$

Where,

A = Cross-sectional area of pile group, at base = $B \times B = B^2$.

P = Perimeter of pile group = 4 B.

r_f = Shear strength of soil = $c = \tau = q_u/2$ (For clayey soils),

$$\therefore Q_{ug} = 4 \, BL \, r_f + B2 \, r_p = 4 \, BL \, c + B_2 \left(9 \, c_p\right).$$

In absence of any other specific data, both c and c_p may be taken equal to $q_u/2$.

If, However, piles of the group are so spaced that they act individually, rather than acting in the group, the total load capacity of n piles is given by,

$$Q_{un} = n\, Q_{up}$$

Where, Q_{up} = Load of individual pile.

The ultimate load (Q_u) of the pile groups will be then equal to lesser of Q_{ug} and Q_{un}, determined above and the permissible load will be equal to Q_u/F.

The pile spacing which utilizes the full capacity of each pile can be found most easily by trial. A spacing of three times the diameter of piles is commonly selected as trial spacing and checked against the criterion that the resistance Q_g obtained by e_q is at least equal to the capacity of single pile multiplied by the number of piles in a group.

The safe load on a group of piles driven through compressible layers to firm material is equal to the number of piles in the group multiplied by the safe load per pile and no reduction need be made on account of the close spacing of piles. However, the frictional resistance in such a case is completely neglected.

4.10 Settlement of Pile Groups

As a rough approximation, the settlement of a group of friction piles can be computed on the assumption that the clay contained between the top of the piles and their lower third point is incompressible and that the load is applied to the soil at this lower third point of the pile.

The presence of pile below this level is ignored. The load is assumed to be uniformly distributed at this level and is assumed to spread at an angle 30° with the vertical.

The soil below this level is divided into a number of layers and the σ_p and $\Delta\sigma$ are calculated at the middle of each layer. The settlement of each layer is calculated from the expression,

$$\rho = \frac{H \times C_c}{1 + e_0} \log_{10} \frac{\sigma_0 + \Delta\sigma}{\sigma_0}$$

Where,

  H = Thickness of the layer.

  e_0 = Initial voids ratio.

  σ_0 = Initial stress at center of the layer = γZ.

Z = Depth of the center of the layer below ground.

$\Delta\sigma$ = Additional stress due to piles = Total pile load Q divided by the area of spread at the center of layer.

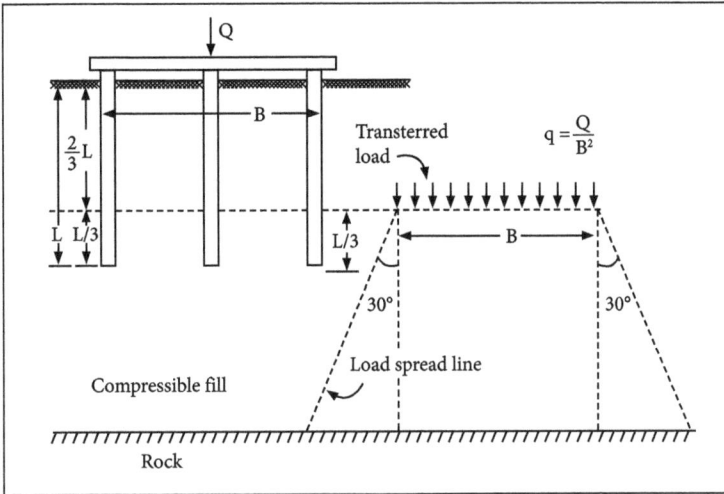

Settlement analysis of pile group in clay.

The total settlement $= \rho_1 + \rho_2 + \rho_3 \ \rho_n$.

Problems

1. For the given design data let us compute the settlement of the group assuming the load to be transferred at 2/3 length of the pile.

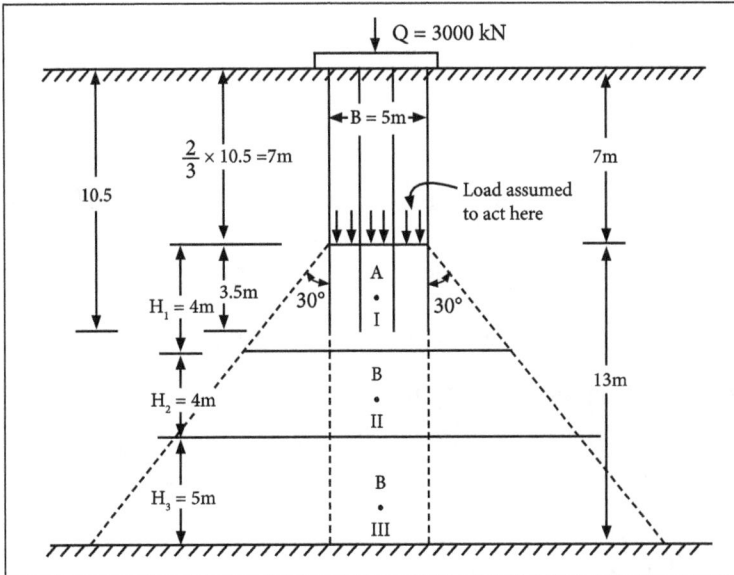

Solution:

Given:

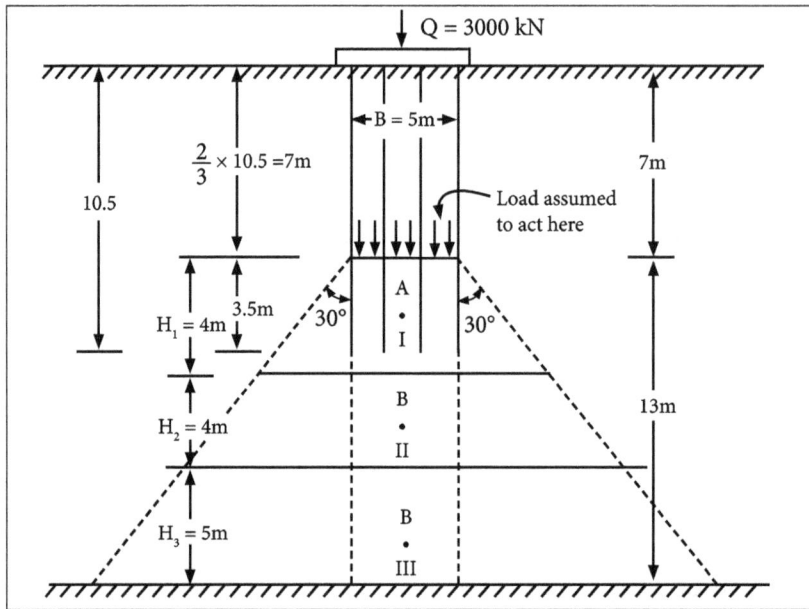

To find: Settlement of the group.

Formula to be used,

$$\rho_1 = \frac{H_1 C_c}{1+e_0} \log_{10} \frac{\sigma_0 + \Delta\sigma}{\sigma_0}$$

$$\text{Area} = \left(B + H_1 \tan 30°\right)^2$$

$$\rho = \rho_1 + \rho_2 + \rho_3$$

The load is assumed to act at $\dfrac{2}{3}L = \dfrac{2}{3} \times 10.5 = 7\,\text{m}$ below ground surface, as shown.

Depth H below the assumed plane of acting of load = 20 - 7 = 13 m.

Assume this depth to be divided into 3 layers of thickness,

$H_1 = 4$ m, $H_2 = 4$ m and $H_3 = 5$ m.

Settlement of First Layer

$$\rho_1 = \frac{H_1 C_c}{1+e_0} \log_{10} \frac{\sigma_0 + \Delta\sigma}{\sigma_0}$$

Here $H_1 = 400$ cm $= 4$ m; $C_c = 0.009 (w_L - 10) = 0.009 (60 - 10) = 0.45$; $e_0 = 1$ (say) σ_0 = initial overburden pressure at the mid-height of the layer,

$$= \gamma Z_1 = 16 \times (7+2) = 144 \text{ kN/m}^2 \left(\text{where } \gamma = 16 \text{ kN/m}^3 \right)$$

Area at the mid-point of first layer,

$$= \left(B + 2 \frac{H_1}{2} \tan 30 \right)^2 = (B + H_1 \tan 30°)^2$$

$\Delta \sigma =$ Pressure due to external load,

$$= \frac{Q}{(B + H_1 \tan 30°)^2} = \frac{3000}{(5 + 4 \times 0.577)^2} = 56.4 \text{ kN/m}^2$$

$$\therefore \rho_1 = \frac{4 \times 0.45}{1+1} \log_{10} \frac{144 + 56.4}{144} = 0.129 \text{ m} = 12.9 \text{ cm} \qquad ...(1)$$

Settlement of the Second Layer

$$\sigma_0 = \gamma Z_2 = 16(7 + 4 + 2) = 208 \text{ kN}/\text{m}^2$$

Area of the mid-point of the layer $= (5 + 2 \times 6 \times 0.577)^2 = 142 \text{ m}^2$

$$\Delta \sigma = \frac{3000}{142} = 21.1 \text{ kN}/\text{m}^2$$

$$\therefore \rho_2 = \frac{4 \times 0.45}{1+1} \log_{10} \frac{208 + 21.1}{208} = 0.038 \text{ m} = 3.8 \text{ cm}. \qquad ...(2)$$

Settlement of the Third Layer

$$\sigma_0 = \gamma Z_3 = 16(7 + 4 + 4 + 2.5) = 280 \text{ kN/m}$$

Area at the mid-point of layer $= (5 + 2 \times 10.5 \times 0.577)2 = 294 \text{ m}^2$

$$\therefore \Delta \sigma = \frac{3000}{294} = 10.2 \text{ kN/m}^2$$

$$\therefore \rho_3 = \frac{5 \times 0.45}{1+1} \log_{10} \frac{280 + 10.2}{280} = 0.017 \text{ m} = 1.7 \text{ cm}$$

Total settlement $\rho = \rho_1 + \rho_2 + \rho_3 = 12.9 + 3.8 + 1.7 = 18.4$ cm.

4.11 Interpretation of Pile Load Test

The pile load test can be performed either on a working pile which forms the foundation of the structure or on a test pile.

The test load is applied with the help of a calibrated jack placed over a rigid square or a circular plate which in turn is placed on the head of the pile projecting above ground level.

The reaction of the jack is borne by a truss or platform which may have gravity loading or alternatively, the truss can be anchored to the ground with the help of anchor piles. In the later case, under-reamed piles or soil anchors can be used for anchoring the truss. Both the arrangements are shown in the below figure.

The load is applied in equal increments of about one-fifth of the estimated permissible load. The settlements are recorded with the help of a three dial gauges of sensitivity 0.02 mm, symmetrically arranged over the test plate and fixed to an independent datum bar. A remote controlled pumping unit can be used for the hydraulic jack. Each load increment is kept for sufficient time till the rate of settlement becomes less than that of 0.02 mm per hour.

The test piles are loaded until ultimate load is reached. Ordinarily, the test load is increased to a value 2½ times the estimated allowable load or to a load which causes a settlement equal to one-tenth of the pile diameter, whichever occurs at the earliest.

The results are plotted in the form of load-settlement curve. The ultimate load is clearly represented by the load settlement curve approaching vertical. If the ultimate load cannot be obtained from the load settlement curve, the allowable load is taken as follows:

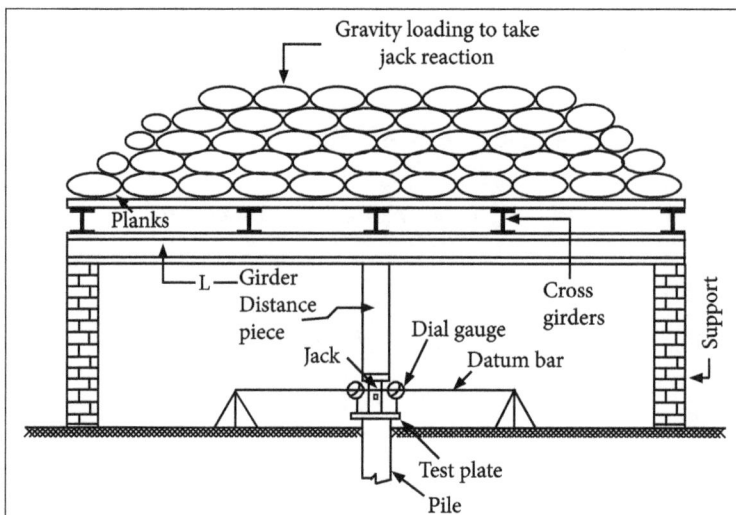

(a) Jack loading: Reaction by loaded platform.

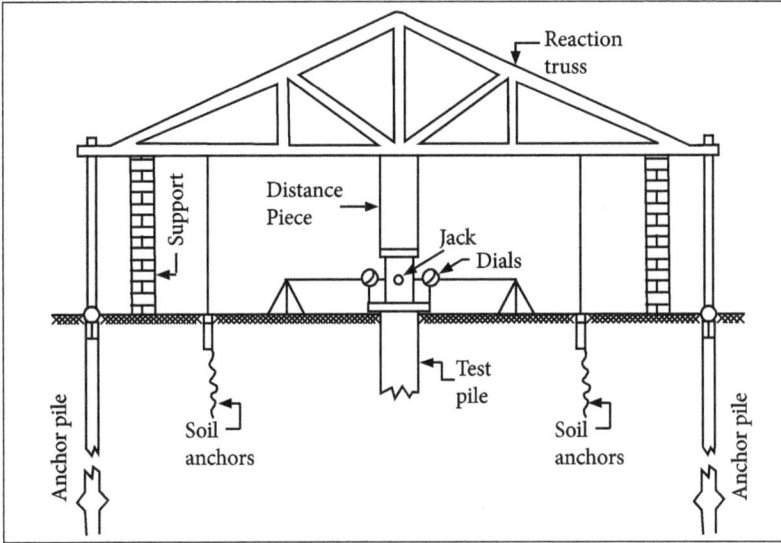

(b) Jack loading: Reaction by anchors.

Arrangements for pile load test:

- One-half to one-third of the final load which causes settlement equal to 10% of the pile diameter.

- Two-thirds of the final load which cause a total settlement of 12 mm.

- Two-thirds of final load which causes a net settlement of 6 mm.

Cyclic Load Test

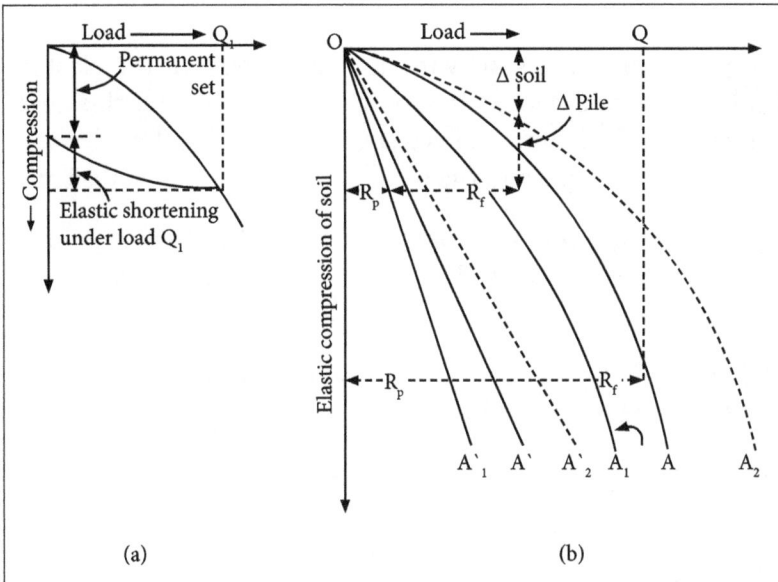

Separation of skin friction and point resistance.

The cyclic load test is particularly useful in separating the load carried by the pile into the skin friction and point bearing resistance. Each load increment is kept on the pile for sufficient time till the settlement decreases to a value less than 0.02 mm per hour.

The load is then completely removed and the elastic rebound of the pile top is measured by means of dial gauges. The next load is then applied and the process repeated.

The cycle of loading and unloading with measurements of settlement and recovery is continued till the final load which causes a marked progressive settlement of the pile is reached.

The elastic compression ΔL of the pile corresponding to any load Q ($= R_f + R_p$) can be calculated from the following expression based on Hooke's law,

$$\Delta L = \frac{(Q - R_f/2)L}{A E}$$

The separation of Q at any stage of loading into R_p and R_f is based on an experimental finding of A.F. Van Weele that the load on the pile toe (i.e. R_p) increases linearly with the elastic compression of soil and that straight line showing the relationship between point resistance and elastic compression of soil is parallel to the straight line portion of the curve drawn between the load on the pile and elastic compression of soil.

The elastic compression of the soil is equal to the total elastic recovery of pile top minus the elastic compression ΔL of the pile. The procedure is described in the following steps:

Steps

- Since R_f is not known to start with, it is assumed that the elastic compression of the pile is zero and hence the elastic compression of soil is equal to the total elastic recovery of pile top. A curve OA_1 [figure (b)] is then drawn between load Q on pile top as abscissa and the elastic compression of soil as ordinate.

- Through origin O, a line OA_1' is drawn parallel to the straight portion of the curve OA. This line is supposed to divide the load into point bearing R_p and skin friction R_f.

- For various loads Q_1, Q_2, Q_3, etc., the skin friction R_{f1} R_{f2}, R_{f3} etc., are determined, as marked in figure (b).

- Corresponding to each value of R_f, the elastic compression of pile is determined from equation The elastic compression of soil is calculated from the relation,

$$\Delta_{soil} = \Delta - \Delta pile$$

Where Δ = total elastic recovery of the pile top (taken from figure (a) from each load):

- Knowing Δ_{soil} for each load Q_1, Q_2, Q_3, etc. a curve is drawn between Q and Δ_{soil}.

- Through the origin O, line OA_2' is drawn parallel to the straight line portion of curve OA_2.

- Step 3, 4, 5 and 6 are repeated to get the final curve and straight line OA' parallel to the straight line portion of curve OA. The third trial of curves gives sufficiently accurate results. From these two, any load Q can be divided to skin friction and point resistance.

The values of skin friction and point resistance corresponding to a load causing a total settlement of one-tenth of the pile diameter are divided by factors of safety of 2 and 2.5 respectively and added together to give the allowable load for the pile.

4.12 Under Reamed Piles

Under reamed piles have mechanically formed enlarged bases that have been as much as 6 m in diameter.

According to the research carried out at Central Building Research Institute Roorkee and elsewhere it is found that under reamed piles provide an ideal solution for foundation in black cotton soil.

The form is that of an inverted cone and can be formed only in stable soils. In such conditions they allow very high load bearing capacities.

The diameter of the pile stem (D) varies from 20 to 50 cm. The diameter of the under-ream bulbs (Du) is generally 2.5 times the diameter of the pile stem. It may however, vary from 2 to 3 times (D) under special circumstances.

In case of double or multi-under-reamed piles, the center to center vertical spacing between two bulbs can vary from 1 ¼ to 1 ½ times the under-reamed diameter (Du). The length of the under-reamed piles varies from 3 to 8 meter and their center to center spacing must generally be not less than 2 times the under-reamed diameter.

Under reamed piles are the most economical and safe foundation in Black cotton soil. Under reamed piles are bored cast in situ concrete piles which has bulb shaped enlargement near base. A Pile having one bulb is called single under reamed pile.

In its closed position, the under reamer fits inside the straight section of a pile shaft and can be expanded at the base of the pile to produce the enlarged base.

The cost advantages of under-reamed piles are due to the reduced pile shaft diameter, resulting in less concrete required to replace the excavated material.

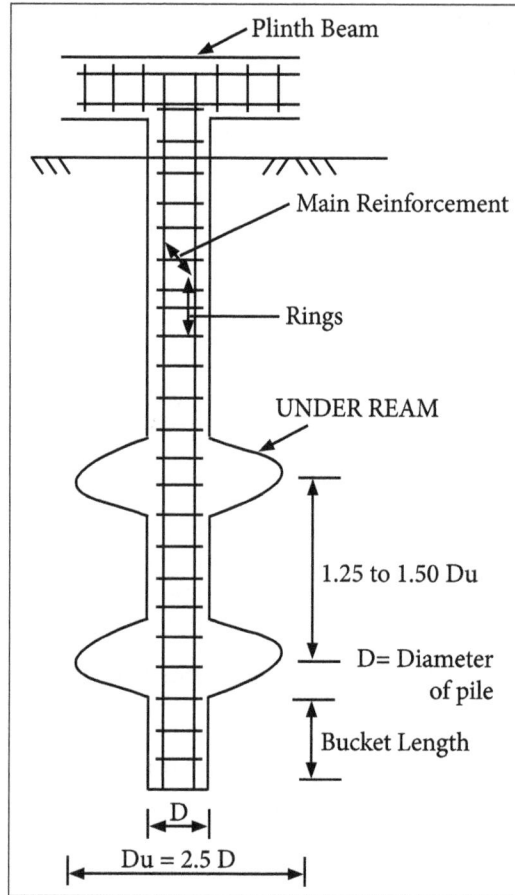

Under reamed pile with two under reams.

Under reamed piles are of two types:

- Pre-cast under reamed piles.

- In-situ cast under reamed piles.

1. Pre-Cast under Reamed Piles: Pre- cast piles needs specialized pile driving equipments. They are advantageous over in-situ cast pile because they do not require holes to be bored and the pile is already cast complete with the reinforcing steel in place.

2. In-Situ Cast under Reamed Piles: This type of pile installation is used to replace soil removed by drilling rather than occupying the space of the displaced soil as in driven piles and thus, it mostly relies on the end-bearing capacity of the earth layer at the drilled depth.

The bearing capacity of the pile is increased by increasing the number of bulbs at the base. Such type of foundations are ideally suitable in the areas where the expansive soil or the black cotton soil is beyond 2.50 meter.

The basic principle of under reamed pile is to anchor the structure at a depth where the ground movements are negligible due to the moisture content variation or other reasons.

Simple tools are needed for the construction of under-reamed piles such as spiral auger, under reaming tool and boring guide. This is a well proven and established technology for the construction of foundation in expansive soils.

For speeding up the construction bore and under ream for deeper pile and large diameter a mechanical rig can be used. The construction and design of such foundation could be done in accordance with Indian Standard Code of Practice IS 2911-Part III.

These piles are ideally suited in soils where considerable ground movements occur because of seasonal variations, filled up grounds or in soft soil strata. The provision of under reamed bulbs has the advantage of increasing the uplift and bearing capacities. It also provides better anchorage even at greater depths. These piles are used efficiently in machine foundations, electrical transmission tower foundation, over bridges and sand water tanks.

Construction of under Reamed Pile Foundation

The various stages involved in the construction of under reamed pile foundation are as under:

- Boring by augers.

- Under-reaming by under-reamer.

- Placing reinforcement cage in position.

- Concreting of pile.

- Concreting of pile caps, plinth beams and curtain walls.

Figure shows various stages in the formation of a under-reamed pile. The equipment required for the construction of pile are (i) auger boring guide (ii) spiral auger with extension rods (iii) under-reamer with soil bucket and (iv) concreting funnel. A portable tripod hoist with winch is required specially for piles larger than 4 to 5 m length and/or of diameter larger than 37.5 cm.

The use of boring guide is essential in order to keep the bore holes vertical and also in position. Each guide is provided with a circular collar and four arms. The collar is fixed to the boring guide on the lower side and it does not allow the mouth of the bore holes to widen due to frequent insertion and removal of the auger and other boring tools.

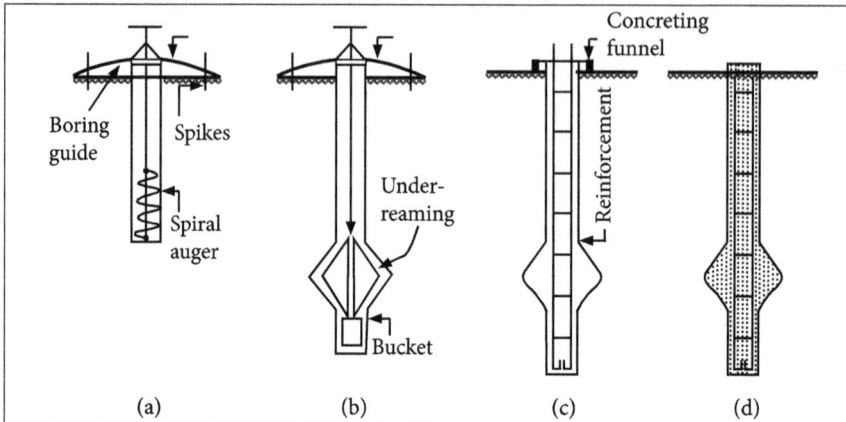

Stages in the construction of under-reamed pile.

After setting the guide assembly in position, the spiral auger is introduced into the circular collar of the guide by opening out the two sets of flaps of the guide assembly. The auger is pressed down and rotated manually until the spirals are half full of earth. The auger is then taken out and earth removed. The auger is again introduced and the boring process is repeated till the required depth is reached.

Under-reaming or enlarging the stem of bore hole at the required depths is achieved by means of the under-reamer consisting of a set of two collapsible blades assembly fixed around the central shaft and a detachable bucket for receiving the cut soil.

The equipment is attached to extension rods and lowered down the hole which has been bored to the required depth, until the bucket rests at the bottom of the bore hole. The guide flaps are then closed. The tool is pressed down constantly and rotated slowly.

The cutting blades of the tool widen out and start cutting the sides of hole. The loose earth is collected in a bucket at the bottom.

When the bucket is full, the assembly is pulled out and bucket is emptied. The depth of the bore hole is checked each time before insertion of the under-reamer so that any loose earth spilled from the bucket is removed. Otherwise, the bucket position will get shifted upwards due to loose soil lying at the bottom. This will shift the position of the bulb.

The under-reamer is then lowered into the bore hole and the process repeated until the cutting blades have expanded fully and no further earth is cut by the blades. Generally, removal of eight buckets full of earth is required for completion of one under-ream.

In the case of double under-reamed pile, further boring is done, after first bulb is formed. After boring to the required depth, the under-reaming for the second bulb is carried out. The dimensions of the bulb can be checked by means of a graduated G.I. pipe assembly.

After the bore and the under-ream are checked, the reinforcement cage, already fabricated, is lowered in the hole. To ensure proper cover for the bars inside the bore hole, 4 Nos. cement mortar discs 2 cm thick and 7.5 cm in diameter with a hole, in the center are inserted into a ring of the reinforcement cage at about the middle height of the pile shaft.

Each reinforcement cage is invariably raised from bottom of the bore hole while concreting to ensure proper cover at bottom. Concreting of the pile is carried out through a concreting funnel placed at the mouth of the bore hole.

As soon as concrete is filled up to top, the funnel is removed and circular steel mould made out of mild steel sheet is placed on the top and the pile shaft for a height of about 15 cm above ground level is converted simultaneously, care being taken to see that the top level of pile shaft is not lower than the bottom of the pile cap to be cast on it. When the concrete in the pile becomes fairly hard, shuttering work for plinth beam and pile caps is started.

4.13 Capacity under Compression and Uplift

Load Test on under-reamed Piles

Piles are usually tested for determining their load carrying capacity in compression, tension and lateral loading. Two categories of tests are conducted: (a) initial tests and (b) routine tests.

Initial tests should be carried out on test piles or working piles, preferably on test piles. In case the initial tests show consistently higher or lower values than the estimated safe allowable loads on piles, designs should be re-examined and necessary modifications made. Routine tests are carried out as check load on working piles.

Procedure for Initial Tests (Compression)

Following are the recommendations of Indian Standard IS: 2911 (Part III): 1973:

The test shall be carried out by applying series of loads to the pile unaided by any other support. Pile groups may be tested as free standing piles or piled foundations as specified:

- The load shall preferably be applied by means of hydraulic jack reacting against a loaded platform or rolled steel joists or suitable load frame held down by soil anchors and piles or other anchorage.

- The anchor piles may also be working piles but they shall be sufficient in number and adequately reinforced to take the full tension with proper factor of safety.

- The reaction available for loading should not be less than 3 times the estimated safe load carrying capacity of the piles.

- The jack should be of adequate capacity preferably with a remote control pump and shall have pressure gauge or other suitable device for reading the applied loads.

Readings of settlement and rebound shall be recorded with the help of at least 3 dial gauges of 0.02 mm sensitivity, positioned at equal distances around the pile. The dial gauges shall be fixed to datum bars resting on non-movable supports at least 5D away from the piles where D is the pile stem diameter.

The test load shall be applied in increments of about 1/5 of the estimated safe load. At each stage of loading/unloading, the load shall be maintained till the rate of movement of the pile top is not more than about 0.02 mm per hour.

Loading shall generally be continued up to 2 1/2 times the estimated safe load or to a settlement of 7.5% of the bulb diameter whichever is earlier.

The safe load on pile shall be the least of the following:

- two-thirds of the final load at which the total settlement attains a value of 12 mm, unless it is established that a total settlement different from 12 mm is permissible in a given case on the basis of nature and type of the structure; in the latter case the actual total settlement permissible shall be used for assessing the safe load instead of 12 mm.

- 50 percent of the final load at which the total settlement equals 7.5 percent on the bulb diameter.

Procedure for Routine Test (Compression)

Loading shall be carried out up to times the allowable load, the reaction provided may be 2 times the allowable load. The procedure followed for the test and determination of the allowable load shall be the same as per initial test excepting clause 5 (ii).

Bored Compaction Piles

Bored compaction piles are the modification of under-reamed pile. These piles are cast-in-situ pile which combines the advantages of both bored and driven piles the method of boring the piles and concreting the pile is the same as that for the under-reamed pile except that the reinforcement cage is not placed in the bore hole before concreting.

After the concreting is over the reinforcement cage is driven through the laid concrete. Due to this feature, the compaction of surrounding soil as well as the concrete are effected and the load carrying capacity is increased by 1.5 to 2 times over normal under-reamed pile. These piles are particularly suitable in loose to medium dense sandy and silty strata.

Also, in case of loose strata, overlying the dense strata specially in submerged soils, these piles can be used with advantage. In such conditions, it is difficult to reach the desired depth in case of bored piles normally without loosening the strata at pile toe.

Figure illustrates the stages of construction of such pile. The bore hole is prepared with the help of spiral auger using guides and then under-reamed. The pile is then concreted without placing the reinforcement cage. The reinforcement cage, enclosing a hollow driving pipe, is then placed over the freshly laid concrete.

A cast iron conical shoe, with iron cleat welded to it, is attached to the reinforcing cage. This assembly of reinforcement cage, hollow pipe and conical shoe, is driven through the freshly laid concrete to full depth by means of suitable drop weight operated with the help of mechanical winches.

The movement of hammer and assembly should be controlled by suitable guiding attachment to ensure vertical penetration of the cage. As the cage is driven into the concrete, soil and concrete gets compacted. This would result in increase in the diameter of the bore hole. Extra concrete is simultaneously poured to keep it level with the ground.

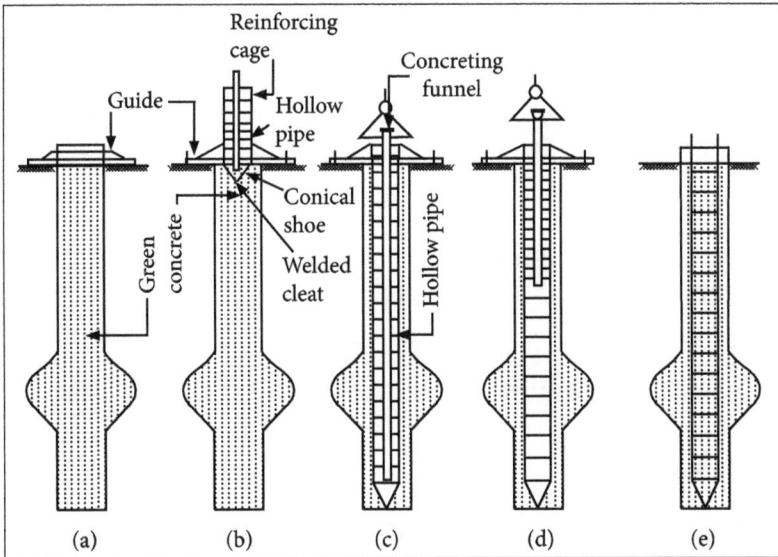

Construction of bored compaction pile.

After driving through the full depth, concrete is filled in drive pipe also. The pipe is then gradually withdrawn leaving the cage, shoe and concrete behind.

Retaining Walls

5.1 Plastic Equilibrium in Soils

Plastic Equilibrium Solution of Bearing Capacity

As the bearing capacity theory assumes, shear failure of the soil mass, the so-called general shear failure, occurs whenever the shear failure zones which propagate throughout the whole subsoil become a continuous shear zone extending from the footing to the surface. Immediately before failure, the soil in the shear failure zone is in a state of plastic equilibrium.

The reason why a plastic equilibrium exists during failure is that failure is apt to occur slowly, i.e. the inertia forces are negligible.

There exists, of course, an equilibrium even before failure. This is not plastic, however. Once the state of plastic equilibrium is attained, plastic flow begins. This is usually concentrated in a narrow band, the shear failure surface which divides the soil mass in two rigid regions, the stationary or immobile subgrade and the moving overburden.

It is assumed that the plastic equilibrium is achieved when Mohr's circle becomes tangential to the failure envelope.

The object of a failure analysis is to find the failure surface and to calculate the stresses which act on this surface. Solutions of the bearing capacity are obtained on the assumption that the soil exhibits plastic behavior which may be expressed with the help of Mohr's failure theory.

In two-dimensional cases, the failure condition is given by,

$$\sqrt{\left[\frac{(\sigma_z - \sigma_x)^2 + 4\tau_{xz}^2}{4}\right]} - \frac{\sigma_z + \sigma_x}{2}\sin\Phi = c\cos\Phi$$

where a_x, a_z are the normal stresses in the direction of the X and Z axes, o is the angle of internal friction and c represents cohesion. In terms of the principal stresses the condition becomes,

$$\frac{\sigma_1 + \sigma_3}{2} - \frac{\sigma_1 + \sigma_3}{2} \sin \Phi = c \cos \Phi$$

$$\frac{\sigma_1 - \sigma_3}{\sigma_1 + \sigma_3 + 2c \cot \Phi} = \sin \Phi$$

The cohesion less soils (c = 0) the failure condition is given by,

$$\sqrt{\left[\frac{\left(\sigma_z - \sigma_x \right)^2 + 4\tau_{xz}^2}{\sigma_z + \sigma_x} \right]} = \sin \Phi$$

Or, in terms of the principal stresses by,

$$\frac{\sigma_1 - \sigma_3}{\sigma_1 + \sigma_3} = \sin \Phi$$

For cohesive perfectly plastic soils ($\varphi = 0$) the condition is,

$$\sqrt{\left[\frac{\left(\sigma_z - \sigma_x \right)^2 + 4\tau_{xz}^2}{4} \right]} = c$$

Or, in terms of the principal stresses,

$$\frac{\sigma_1 - \sigma_3}{2} = c$$

A rigorous solution of the state of plastic equilibrium is defined by the system of differential equations of equilibrium, the appropriate boundary conditions of loads and the shear failure condition selected from to fit the case in question.

Generally, infinitely many states of plastic equilibrium are possible. The true state is that which yields the highest failure load and simultaneously satisfies the equation of compatibility.

Any state of plastic equilibrium provides a statical lower bound and a kinematical upper bound of the true collapse load. In practice, the bounds are frequently so close to the exact collapse load that they may be used as an approximation.

The rigorous solutions are presented by Harr who, according to Sokolovsky, makes use of a system of non-linear differential equations, which allow us to solve variable and inclined loads acting on the surface.

The assumption of plastic behaviour introduced in the form of the failure condition expressed in terms of stresses, represents merely a simplified formulation of the problem.

This formulation neglects the inelastic volume dilatancy which accompanies plastic shear of all soils, the dependence of dilatancy on the confining stress, strain-hardening and strain-softening after the attainment of the peak stress.

An entirely new formulation which takes these phenomena into account and offers a much more accurate description of the real response of soils was recently developed as the so-called endochronic theory.

The plastic behaviour and mohrs envelope may be obtained as special cases of this formulation. In practical applications to predications of the bearing capacity, the endo-chronic theory requires a finite element computer analysis.

5.2   Active and Passive States

When the wall moves away from the backfill, there is a decrease in the pressure on the wall and this decrease continues until a minimum value is reached. After which there is no reduction in the pressure and the value will become constant. This kind of pressure is referred as active earth pressure.

Passive Earth Pressure

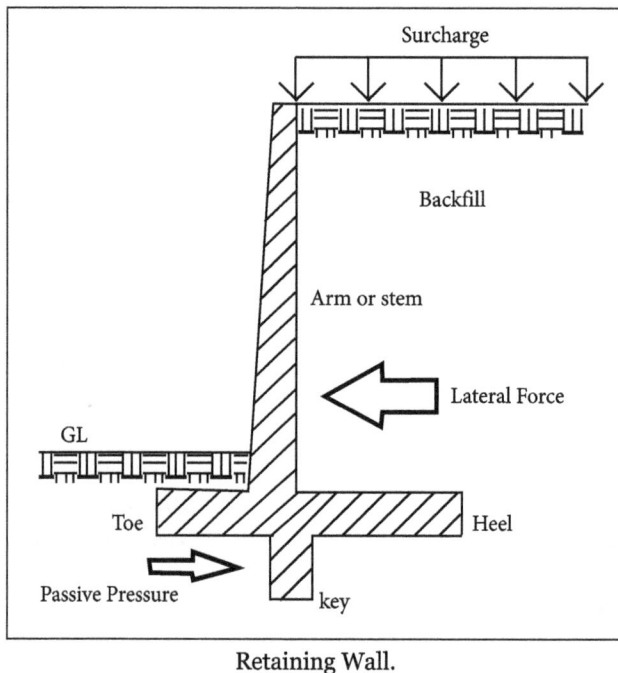

Retaining Wall.

When the wall moves towards the backfill, there is an increase in the pressure on the wall and this increase continues until a maximum value is reached. After which there is no increase in the pressure and the value will become constant. This kind of pressure is referred as passive earth pressure. When the wall is about to slip due to lateral thrust from the backfill, a resistive force is applied by the soil in front of the wall.

5.3   Rankine's Theory

Rankine's Theory

The Rankine's theory assumes that:

- There is no wall friction (δ = 0).

- The ground and failure surfaces are straight planes.

- The resultant force acts parallel to the backfill slope.

In case of retaining structures, the earth retained may be filled up in earth or natural soil.

These backfill materials may exert some lateral pressure on the wall. If the wall is rigid and does not move with the pressure exerted on the wall, the soil behind the wall will be in a state of elastic equilibrium. Consider the prismatic element E in the backfill at depth Z.

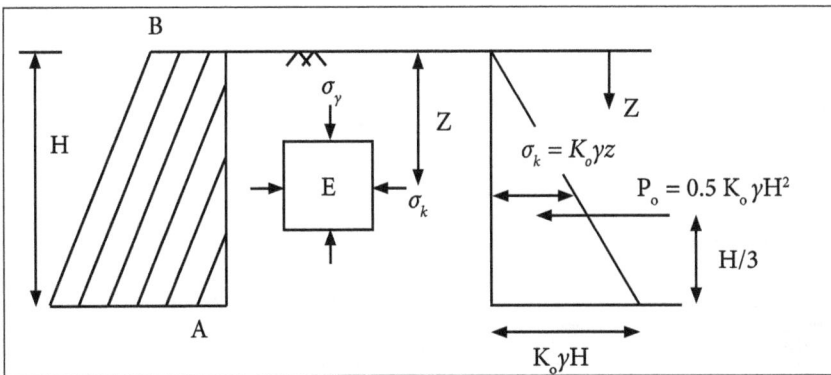

Lateral earth pressure for at rest condition.

The element E is subjected to the following pressures,

Vertical pressure = σ_v = γz

Lateral pressure = σ_k

If we consider the backfill is homogeneous then both the σ_v and σ_h increases rapidly with depth z.

In that case the ratio of vertical and lateral pressures remain constant with respect to depth, that is $\sigma_h/\sigma_v = \sigma_h/\gamma_z$ = constant = K_o.

Where, K_o is the coefficient of earth pressure at rest condition.

At Rest Earth Pressure

The at-rest earth pressure coefficient (K_o) is applicable for determining the active pressure in clay for strutted systems.

Because of the cohesive property of clay there will be no lateral pressure exerted in the at rest condition up to some height when the excavation is made with time, creep and swelling of the clay will occur and a lateral pressure will develop.

This coefficient takes the characteristics of clay into account and will always give a positive lateral pressure. This lateral earth pressure acting on the wall of height H may be expressed as $\sigma_h = K_o\gamma H$.

The total pressure for the soil at rest condition, $P_o = 0.5\ K_o\gamma H^2$.

The value of K_o depends on the:

- Relative density of sand.

- The process by which the deposit was formed.

If this process does not involve artificial tamping the value of K_o ranges from 0.4 for loose sand to 0.6 for dense sand.

Tamping of the layers may increase it up to 0.8.

From elastic theory, $K_o = \mu/(1-\mu)$, where μ is the Poisson's ratio.

According to Jaky, a good approximation of Ko is given by, $K_o = 1 - \sin \varphi$.

Different Values of Ko

Soil type	Typical value for Poisson's ratio	Ko
Clay, saturated	0.40 - 0.50	0.67 - 1.00
Clay, unsaturated	0.10 - 0.30	0.11 - 0.42
Sandy clay	0.20 - 0.30	0.25 - 0.42
Silt	0.30 - 0.35	0.42 - 0.54
Sand - Dense - Course (valid up to 0.4 - 0.7) - Fine-grained (valid up to 0.4 - 0.7)	0.20 - 0.40	0.25 - 0.67
	0.15	0.18
	0.25	0.33
Rock	0.10 - 0.40	0.11 - 0.67

Rankine's earth pressure against a vertical section with the surface horizontal with cohesion less backfill.

Active Earth Pressure

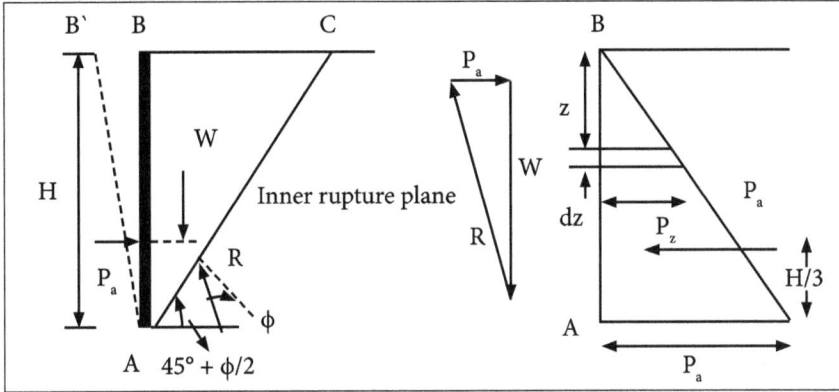

Rankine's active earth pressure in cohesionless soil.

The lateral pressure acting against a smooth wall AB is due to mass of soil ABC above the rupture line AC which makes an angle of ($45° = \varphi/2$) with the horizontal.

The lateral pressure distribution on the wall AB of height H increases in same proportion to depth.

The pressure acts normal to the wall AB.

The lateral active earth pressure at A is $P_a = K_A \gamma H$ which acts at a height H/3 above the base of the wall.

The total pressure on AB is therefore calculated as follows,

$$P_a = \int_0^H p_z dz = \int_0^H K_A \gamma z\, dz = 0.5\, K_A \gamma H^2$$

Where,

$$K_A = \tan^2\left(45° + (\phi/2)\right)$$

Passive Earth Pressure

If the wall AB is pushed into the mass to such an extent so as to impart uniform compression throughout the mass, the soil wedge ABC in figure will be in Rankine's Passive State of plastic equilibrium.

The inner rupture plane AC makes an angle ($45° = \varphi/2$) with the vertical AB.

The pressure distribution on the wall is linear as shown.

The lateral passive earth pressure at A is $P_p = K_p \gamma H$. which acts at a height $H/3$ above the base of the wall.

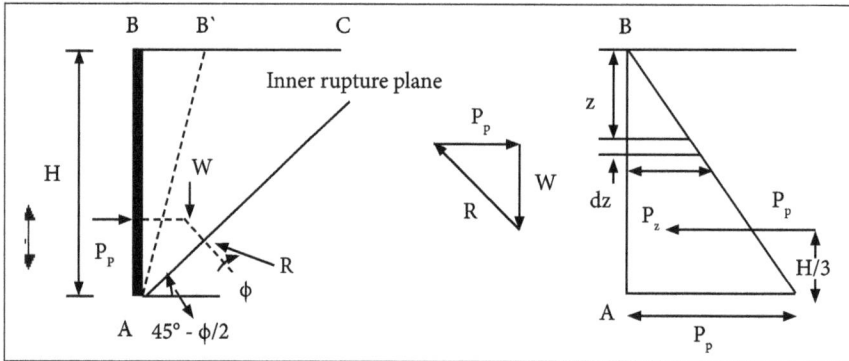

Rankine's passive earth pressure in cohesionless soil.

The total pressure on AB is therefore,

$$P_p = \int_0^H p_z \, dz = \int_0^H K_p \gamma z \, dz = 0.5 K_p \gamma H^2,$$

Where,

$$K_p = \tan^2\left(45° + (\phi/2)\right)$$

Rankine's Active Earth Pressure with a Sloping Cohesion Less Backfill Surface

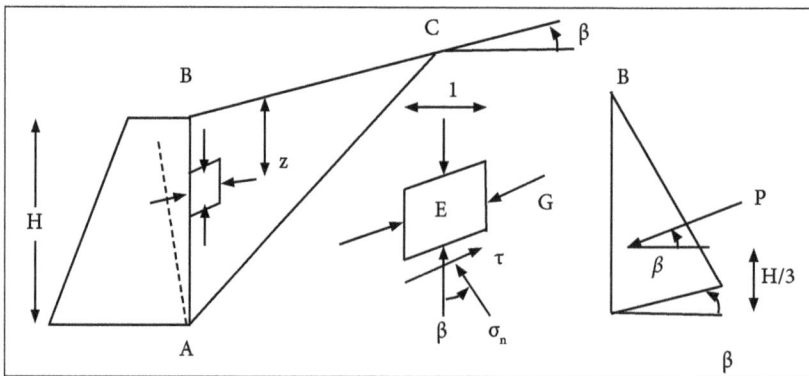

Rankine's active pressure for a sloping cohesionless backfill.

Figure shows a smooth vertical gravity wall with a sloping backfill with cohesion less soil.

As in the case of the horizontal backfill, active case of plastic equilibrium can be developed in the backfill by rotating the wall about A away from the backfill.

Let AC be the plane of rupture and the soil in the wedge ABC is in the state of plastic equilibrium.

The pressure distribution on the wall is shown in figure.

The active earth pressure at depth H is $P_a = K_A \gamma H$ which acts parallel to the surface.

The total pressure per unit length of the wall is $P_a = 0.5 K_A \gamma H^2$ which acts at a height of H/3 from the base of the wall and parallel to the sloping surface of the backfill.

In case of active pressure,

$$K_A = \cos\beta \left(\cos\beta - \sqrt{(\cos^2\beta - \cos^2\phi)}\right) / \left(\cos\beta + \sqrt{\cos^2\beta - \cos^2\phi}\right).$$

In case of passive pressure,

$$K_p = \cos\beta \left(\cos\beta + \sqrt{(\cos^2\beta - \cos^2\phi)}\right) / \left(\cos\beta - \sqrt{\cos^2\beta - \cos^2\phi}\right).$$

Rankine's Active Earth Pressures of Cohesive Soils with Horizontal Backfill on Smooth Vertical Walls

In case of cohesion less soils, the active earth pressure at any depth is given by $P_a = K_A \gamma z$.

In case of cohesive soils the cohesion component is included and it can be expressed as,

$$P_a = K_A \gamma z - 2c \sqrt{K_A}$$

When,

$$P_a = 0, z = z_0 = \left(2_c \sqrt{K_A}\right) / \gamma$$

This depth is known as the depth of tensile crack.

Assuming that the compressive force balances the tensile force (-), the total depth where tensile and compressive force neutralizes each other is $2z_0$.

This is the depth up to which a soil can stand without any support and is sometimes referred as the depth of critical depth or vertical crack $\left(H_c\right)\left(H_c = 4c \sqrt{K_A}\right) / \gamma$.

However Terzaghi from field analysis observed that $\left(H_c = 4c \sqrt{K_A}\right) / \gamma - z_0$.

Where, $z_0 \approx H_c/2$ and is not more than that.

The Rankine formula for passive pressure can be used correctly only when the embankment slope angle is negative or equals zero.

If a large wall friction value develops, the Rankine Theory is not correct and will give less conservative results. This theory is not intended to be used for determining earth pressures directly against a wall.

This theory is intended to be used for determining earth pressures on a vertical plane within a mass of soil.

5.4   Cohesion Less and Cohesive Soil

Cohesive Soil

It is hard to break up when dry and exhibits significant cohesion when submerged. It includes clayey silt, sandy clay, silty clay, clay and organic clay. "Dry soil" means a soil that does not exhibit visible signs of moisture content.

Cohesion Less Soil

It is free-running type of soil, such as sand or gravel, whose strength depends on friction between particles. It is also called as frictional soil.

5.5   Coulomb's Wedge Theory

The assumptions are:

- The soil is isotropic and homogeneous.
- The surface of rupture is a plane.
- The failure wedge is a rigid body.
- There is friction between the walls.
- Back of wall need not be vertical.
- The soil is cohesion less.
- Coulomb's equation of shear strength is valid.
- Coulomb made his derivation based on the limit equilibrium approach.

In Active Case

Below figure shows the cross section of a retaining Wall.

The equilibrium analysis of failure wedge ABC involves:

- Weight of wedge ABC (magnitude and direction known).

- R (direction known, magnitude unknown).

- P_a (direction known, magnitude unknown).

Hence, from the triangle of forces P_a can be determined.

Coulomb's Active Earth Pressure.

Weight of wedge ABC:

From $\triangle$ ABC,

Area of $\triangle$ ABC = ½ AD × BC

$$BC / AB = \left\{ \sin(\alpha+\beta) / \sin(\theta-\beta) \right\}$$

$$BC = AB \left\{ \sin(\alpha+\beta) / \sin(\theta-\beta) \right\}$$

Again,

$$AD = AB \sin\left[180^0 - (\alpha+\theta) \right] = AB \sin(\alpha+\theta)$$

$$\text{Area of } \Delta \text{ ABC} = \frac{1}{2} \text{ AB } \sin(\alpha+\theta) \times \text{AB } \frac{\sin(\alpha+\beta)}{\sin(\theta-\beta)}$$

$$\Delta \text{ ABC} = \frac{H^2}{2\sin^2 a}\left[\sin(\alpha+\theta)\frac{\sin(\alpha+\beta)}{\sin(\theta-\beta)}\right]$$

$$W = \gamma \times V = \frac{\gamma H^2}{2\sin^2 a}\left[\sin(\alpha+\theta)\frac{\sin(\alpha+\beta)}{\sin(\theta-\beta)}\right]$$

Triangle of Forces for W, P_a and R:

From the sine rule,

$$\frac{P_a}{\sin(\theta-\phi)} = \frac{W}{\sin\left[180°-\{(\theta-\phi)+(\alpha-\delta)\}\right]}$$

Substituting the value of W in the equation we get,

$$P_a = \frac{\gamma H^2}{2\sin^2 \alpha}\left\{\sin(\alpha+\theta)\frac{\sin(\alpha+\beta)}{\sin(\theta-\beta)}\right\}\frac{\sin(\theta-\phi)}{\sin(180°-\theta+\phi-\alpha+\delta)}$$

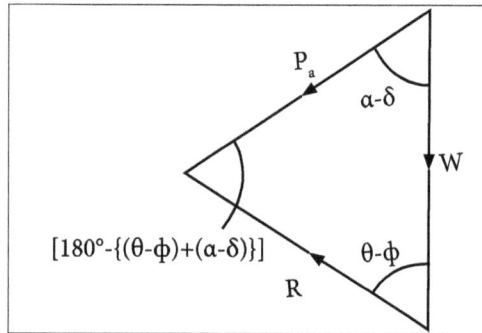

In order to obtain a maximum value of P_a,

$$\frac{\partial P_a}{\partial \theta} = 0$$

$$P_a = \frac{\gamma H^2}{2}\left[\frac{\sin^2(\phi+\alpha)}{\sin^2 \alpha \sin(\alpha-\delta)\left\{1+\sqrt{\frac{\sin(\phi+\delta)\sin(\phi-\beta)}{\sin(\alpha-\delta)\sin(\alpha+\beta)}}\right\}}\right]$$

$$P_a = K_a \frac{\gamma H^2}{2}$$

Where,

$$K_a = \left[\frac{\sin^2(\phi+\alpha)}{\sin^2\alpha\sin(\alpha-\delta)\left\{1+\sqrt{\dfrac{\sin(\phi+\delta)\sin(\phi-\beta)}{\sin(\alpha-\delta)\sin(\alpha+\beta)}}\right\}} \right]$$

5.6   Condition for Critical Failure Plane

The favorable conditions of plane failure are as follows:

- The dip of the planar discontinuity should be less than the dip of the slope face of the dip direction of the slope face.

- The dip direction of the planar discontinuity should be within (± 20°).

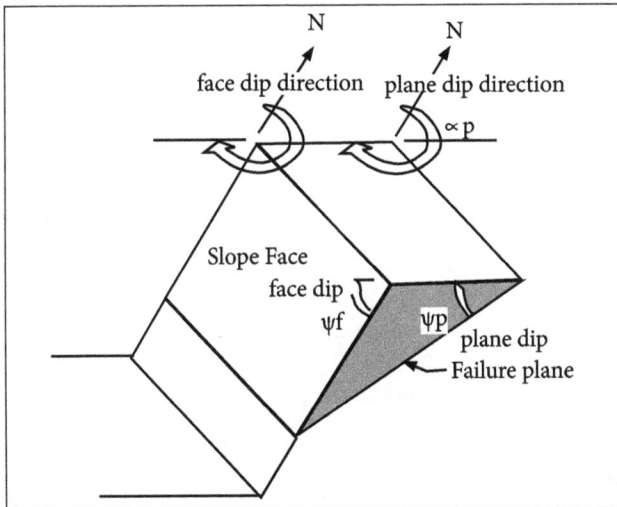

Plane failure with condition of failures.

- The dip of the planar discontinuity must be greater than the angle of friction of the surface.

5.7   Earth Pressure on Retaining Walls of Simple Configurations

The earth pressures are defined as follows: Let us consider a retaining wall with a plane vertical face, as shown in figure, which is backfilled with cohesion less soil. If the wall does not move even after filling the materials then the pressure exerted on the wall is referred as pressure for the at rest condition of the wall.

If suppose the wall rotates gradually about the point A and moves away from the backfill, the unit pressure on the wall gradually gets reduced and after a certain displacement of the wall at the top, the pressure reaches a constant value. This pressure is the minimum possible pressure. The pressure is known as the active earth pressure since the weight of the backfill is responsible for the movement of the wall. If the wall is smooth, the resultant pressure acts normal to the face of the wall.

If the wall is rough, the resultant pressure acts at an angle δ which is the normal to the face. This angle δ is known as the angle of wall friction. When the wall moves away from the backfill, the soil also tends to move forward. If the wall movement is sufficient, a soil mass of weight W ruptures along the surface AC'C shown in figure. This surface is slightly curved. If the surface is assumed to be plane surface AC the analysis would indicate that this surface would make an angle of 45°+ φ/2 with the horizontal.

If the wall is now rotated about A towards the backfill, the actual failure plane ACC is also a curved surface. However, if the failure surface is approximated to a plane AC, this makes an angle 45°+ φ/2 with the horizontal and the pressure on the wall increases from the value of at rest condition to a maximum possible value.

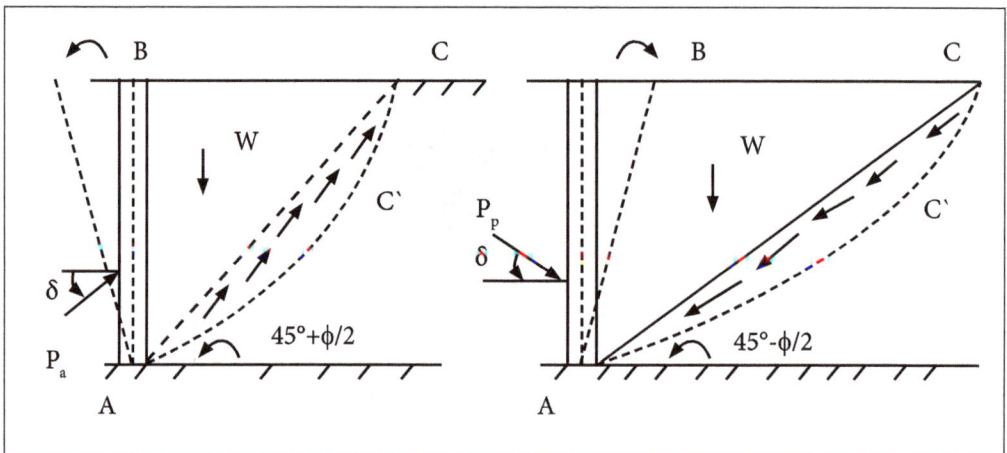

(a) Active earth pressure (b) Passive earth pressure.
Wall movement for the development of active and passive earth.

The maximum pressure p_p that is developed is termed as passive earth pressure. The pressure is known as passive earth pressure since the weight of the backfill opposes the movement of the wall. If the wall is rough, it makes an angle of δ with the normal. The gradual increase or decrease of the pressure of the wall with the movement of the wall from the "at rest condition" can be depicted as shown in figure. The movement Δp, required to develop passive state is considerably larger than the Δa required for active case.

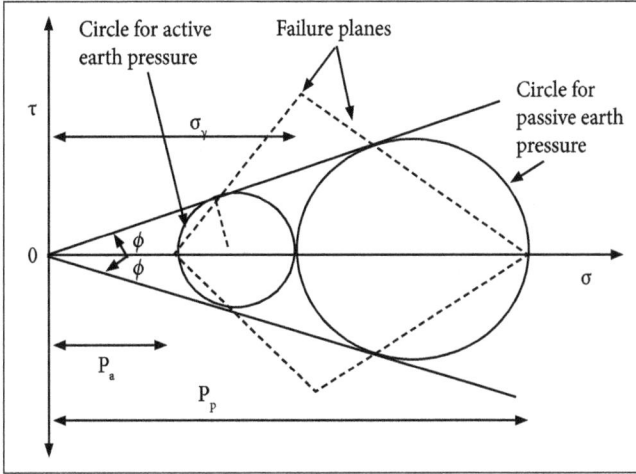

Mohr stress diagram.

5.8   Culmann Graphical Method

Active Case

In this method the retaining wall is drawn to a suitable scale.

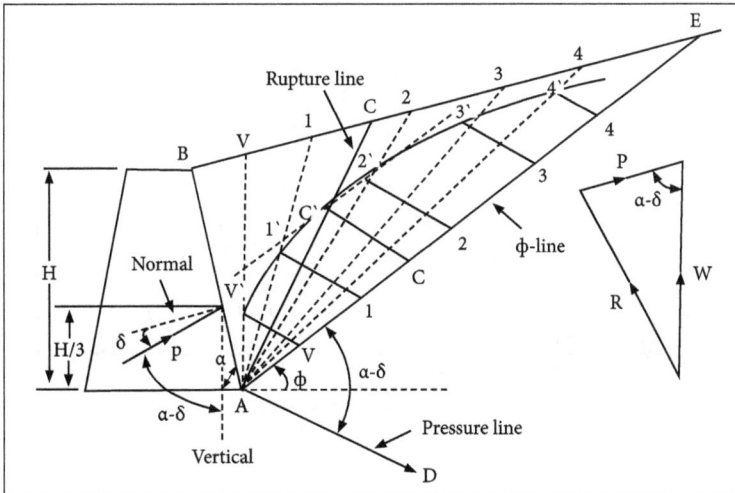

Culmann's Graphical Construction Active Case.

The Various steps in the procedure are:

- Draw ϕ-line AE at an angle ϕ with the horizontal.

- Lay off on AE distances AV, A_1, A_2, A_3 etc. to a suitable scale to indicate the weight of wedges ABV, AB_1, AB_2, AB_3 and so on.

- Lay off AD at an angle which is equal to (α- δ) to the line AE. The line AD is called pressure line.

- Draw lines parallel to AD from points V, 1, 2, 3 to intersect the assumed lines AV, A_1, A_2, A_3 at points V', 1', 2', 3' etc. respectively.

- Join points V', 1', 2', 3' etc. by a smooth curve which is the pressure locus.

- Select the point C' on pressure locus such that the tangent to the curve is parallel to the φ-line AE.

- Draw CC' parallel to the pressure line AD. The magnitude of CC' in its natural units gives the active pressure Pa.

- Join AC' and produce to meet the surface of the backfill at C. AC is the rupture line.

Passive Case

In this method a retaining wall is drawn to a suitable scale.

Culmann's Graphical Construction Passive Case.

The various steps in the procedure are:

- Draw φ-line AE at an angle φ below the horizontal.

- Lay off on AE distances A_2, A_3, A_4 etc. to a suitable scale to indicate the weight of wedges AB_2, AB_3, AB_4 and so on.

- Lay off AD at an angle equal to $(\alpha + \delta)$ to the line AE. The line AD is referred as pressure line.

- Draw lines parallel to AD from points, 2, 3, 4 etc. to intersect the weight vectors A_2, A_3, A_4 at points 2', 3', 4' etc. respectively.

- Join points, 2', 3', 4' etc. by a smooth curve which is the pressure locus.

- Select the point C' on pressure locus curve such that the line tangent to the curve is parallel to ϕ-line AE.

- Draw CC' parallel to the pressure line AD. The magnitude of CC' in its natural units gives the passive pressure Pp.

- Join AC'. The line cuts the surface of the backfill at C. The line AC is the rupture line.

5.9 Pressure on the Wall Due to Line Load

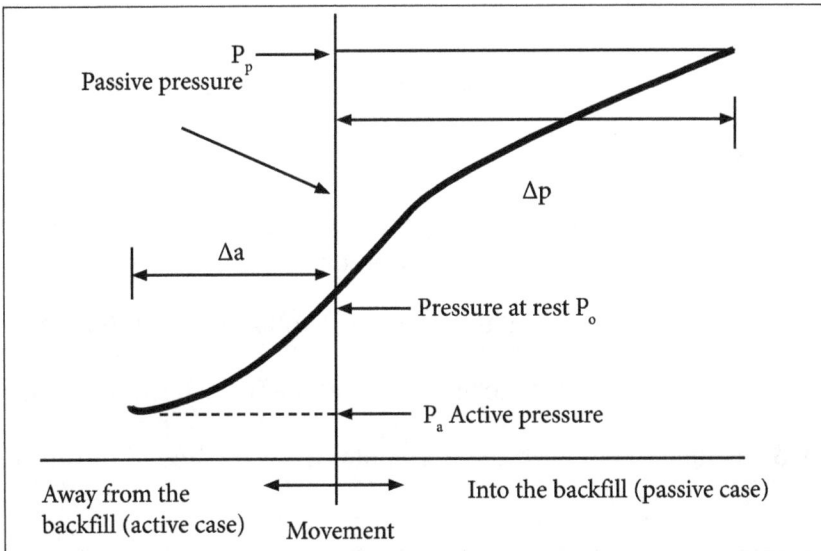

Development of active and passive earth pressures.

Lateral Earth Pressure For At Rest Condition: The earth retained behind retaining walls may be natural earth or filled up soil. These backfill materials can exert certain lateral pressures on the wall. If the wall is rigid and does not move with the pressure exerted on the wall, the soil behind the wall will be in a state of elastic equilibrium.

To reach the plastic state, for p_a to develop we need a minimum displacement p_a. If the displacement δ_a developed is less than δ_a, the pressure developed is called partially mobilized earth pressure which is more than p_a.

Similarly for passive case the wall moves towards the soil and the soil tries to resist the movement of the wall. To reach the passive case a very high movement is required, δ_p. When we are measuring the passive pressure we are measuring the resistance of the soil against the wall movement.

Hence for all practical purposes we get fully mobilized active earth pressure but partially mobilized passive earth pressure as such a large movement does not occur usually. Hence it is more acceptable to design with active earth pressure.

Thus when δ_a is reached active pressure is fully mobilized but passive pressure is partially mobilized. Therefore we should design with (p_a - a small % of p_p), as, we are making an unsafe design when we are designing with (p_a - fully mobilized p_y) and an uneconomic one when we are designing with only p_a.

To get fully mobilized active and passive earth pressures the following wall movements are required approximately,

   δ_a = (0.1 - 0.4)% of the height of the wall.

   δ_p = (5-10)% of the height of the wall.

For dense sand p_a mobilized at 0.1% strain and p_p at 5% strain.

For loose sand p_a mobilized at 0.4% strain and pp at 10% strain.

Rebhann's Methods

Rebhann (1871) is attributable with having presented the criterion for the direct location of the failure plane assumed within the Coulomb's theory.

The steps concerned in the graphical method are as follows, with reference to figure:

- Let AB represent the back face of the wall and AD the backfill surface.

- Draw BD φ inclined at with the horizontal from the heel B of the wall to meet the backfill surface in D.

- Draw BK inclined at Ψ, (= α − δ) with BD, that is the Ψ line.

- Through A, draw AE parallel to the 4-line to meet BD in E. (Alternatively, draw AE at φ + δ) with AB to meet BD in E).

- Describe a semi-circle on BD as diameter.

- Erect a perpendicular to BD at E to meet the semi-circle in F.

- With B as center and BF as radius draw an arc to meet BD in G.

- Through G, draw a parallel to the Ψ-line to meet AD in C.

- With Gas center and GC as radius draw an arc to cut BD in L; join CL and additionally draw a perpendicular CM from C on to LG.

- BC is the needed rupture surface.

The active thrust P_a is given by,

$$P_a = \frac{1}{2} \gamma x^2 . \sin \psi,$$

Where,

$$CG = LG = x$$

$$= \gamma . (\Delta \, CGL)$$

$$= - \gamma . x . n.$$

Where n = CM, the altitude on the LG.

Earth Pressure Distribution

Here $N\phi = \tan^2 (45° + \phi / 2)$, called flow value.

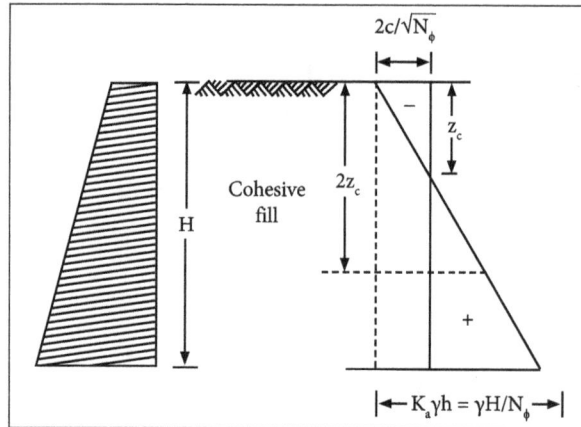

Active pressure distribution for a cohesive soil.

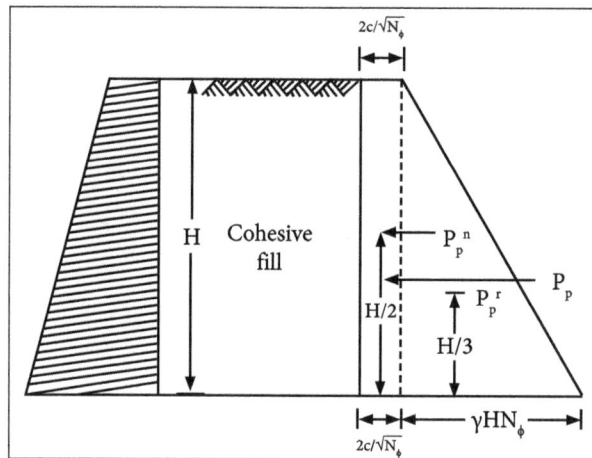

Passive pressure distribution for a cohesive soil.

5.10 Stability Analysis of Retaining Walls

Stability of the Slopes of Earth Dam

Earth dams should be safe against slope and foundation failure for all operational conditions. There are three generally recognized crucial stages based on pore pressure that the stability of the embankment should be determined.

These three situations are:

- Steady-state seepage.

- End of construction.

- Rapid drawdown.

Usually construction pore pressure reaches their most values once the embankment reaches maximum height. After the reservoir has been filled for a long time, pore pressure can be evaluated by steady state seepage conditions and may be calculated by the construction of flow net. Rapid lowering of the reservoir produces a third critical situation, particularly for low permeable soils.

The upstream slope stability will be crucial for the construction of rapid drawdown condition. The downstream slope ought to be checked for the construction and steady state seepage condition:

- Check factor of safety against sliding.

- Check soil bearing pressure.

- Check factor of safety against overturning.

Lateral earth pressure is generally calculated based on Coulomb's or Rankine theories. Lateral earth pressure is assumed to be distributed triangularly. The location of resultant is at 1/3 of height. If there is surcharge, the lateral earth pressure from the surcharge is distributed uniformly. The resultant is at ½ of height. The lateral earth pressure is determined at the edge of heel.

The rotating point for overturning is normally assumed at the bottom of toe. The height of the soil that is used to calculate the lateral earth pressure should be from top of earth to the bottom of footing.

Elements that resisting and overturning are weight of stem, weight of footing, weight of the soil above footing. If there is a surcharge then the weight of surcharge can also be considered.

The factor of safety against the overturning is given by the resisting moment divided by the overturning moment. Acceptable factor of safety is in the range of 1.5 to 2.

The driving force that causes the retaining wall to slide is the lateral earth pressure from the soil and the surcharge. The forces that resist sliding are passive pressure at the toe, the friction at the base of the footing and the passive pressure against the key if used.

The factor of safety against sliding is the total resisting force divided by the total driving force. Acceptable factor of safety is between 1.5 to 2.

Overturning

The factor of safety against overturning can be determined by the following methods.

1. The overturning moment is calculated as,

$$M_o = \frac{1}{6}\gamma K_a H^3 + \frac{1}{2}q K_a H^2$$

Where,

K$_a$ is active pressure coefficient.

g is unit weight of soil.

q is surcharge.

H is the height from top of earth to bottom of footing.

2. The resisting moment can be determined as follows,

$$M_R = W_s X_s + W_f X_f + W_e X_e + W_k X_k + W_q X_q$$

Where, W_e, W_f, W_k, W_s, W_q are weight of earth, footing, key, stem and surcharge, X_e, X_f, X_k, X_s, X_q are distance from the center of the earth, footing, key, stem and surcharge to the rotation point at toe.

3. The factor of safety against overturning can be calculated as,

$$FS = \frac{M_R}{M_o}$$

Bearing Pressure

The bearing pressure is calculated by the following methods:

1. The center of the total weight from the edge of toe is,

$$X = \frac{M_R - M_o}{W}$$

Where, W is the total weight of the retaining wall including the footing, stem, earth and surcharge.

2. The eccentricity, e = B/2-X.

3. If e < B/6, the maximum and minimum footing pressure can be determined as,

$$Q_{max} = \frac{W}{B}\left(1 + \frac{6 \times e}{B}\right)$$

$$Q_{min} = \frac{W}{B}\left(1 - \frac{6 \times e}{B}\right)$$

Where,

Q_{max}, Q_{min} are maximum and minimum footing pressure,

B is the width of the footing.

4. If $e > B/6$, Q_{min} is zero,

$$Q_{max} = \frac{2 \times W}{3\left(\dfrac{B}{2} - e\right)}$$

5. Q_{max} should be less than allowable soil bearing capacity of footing soil.

Sliding

1. The driving force for sliding can be calculated as,

$$P_h = \frac{1}{2} \gamma K_a H^2 + q K_a H$$

2. The friction resisting force at the base of footing is calculated as,

$$F_R = \mu W$$

Where,

> f is internal friction of the soil.

> m is friction coefficient between concrete and soil.

> m is often taken as tan (2/3 f).

3. The passive resistance at the toe of retaining wall is calculated as,

$$P_p = \frac{1}{2} \gamma K_p h^2$$

Where,

> h is the height from top of soil to bottom of footing at toe.

> K_p is passive earth pressure coefficient.

If a key is used to help resist sliding then h is the height from top of soil to the bottom of the key.

4. The factor of safety is calculated as:

$$FS = \frac{F_R + P_p + P_k}{P_h + P_h'}$$

Problem

1. Let us design cantilever retaining wall with horizontal backfill & no surcharge with the given data: Height of stem: 10 ft, Thickness of stem: 1 ft, Thickness of footing: 1 ft, Width of footing: B = 6 ft, Length of heel: 4 ft, Height of soil above heel: 10 ft, Length of toe: 1 ft, Height of soil above toe: 1 ft, Unit weight of backfill soil: g = 115 pcf, Internal friction angle of soil: 30°, Allowable soil bearing capacity for footing soil: 3000 psf, Friction coefficient between concrete and soil: 0.5.

Solution:

Given:

Height of stem: 10 ft.

Thickness of footing: 1 ft.

Width of footing: B = 6 ft.

Thickness of stem: 1 ft.

Length of heel: 4 ft.

Length of toe: 1 ft.

Height of soil above heel: 10 ft.

Height of soil above toe: 1 ft.

Unit weight of backfill soil: g = 115 pcf.

Internal friction angle of soil: 30°.

Allowable soil bearing capacity for footing soil: 3000 psf.

Friction coefficient between concrete and soil: 0.5.

Formula to be used:

$$K_a = \tan\left(45 - f/2\right)^2$$

$$M_o = g\, K_a\, H^3/6$$

$$F_S = M_R / M_o$$

$$X = \left(M_R - M_o\right)/W$$

$$Q_{max} = \frac{W}{B}\left(1 + \frac{6.e}{B}\right)$$

$$Q_{min} = \frac{W}{B}\left(1 - \frac{6.e}{B}\right)$$

$$P_h = g\, K_a\, H^2 / 2$$

$$F_S = (F_r + P_p)/P_h$$

$$P_p = g\, K_p\, H^2 / 2$$

Requirements

Check stability against overturning and sliding and soil bearing capacity:

1. Check Overturning Stability

Rankine's active earth coefficient: $K_a = \tan(45 - f/2)^2 = 0.333$.

Height from top of backfill soil to bottom of footing is given by: $H = 10 + 1 = 11$ ft.

Consider one-foot width of soil.

Overturning moment: $M_o = g\, K_a\, H^3/6 = 8504$ ft-lb.

Calculate Resisting Moment

Weight of stem: $W_s = 150 \times 10 \times 1 = 1500$ lbs.

Distance from center of stem to edge of toe: $X_s = 1.5$ ft.

Weight of footing: $W_f = 150 \times 6 \times 1 = 900$ lbs.

Distance from center of footing to edge of toe: $X_f = 3$ ft.

Weight of earth: $W_e = 115 \times 10 \times 4 = 4600$ lbs.

Distance from center of earth to edge of toe: $X_e = 4$ ft.

Resisting moment: $M_R = 1500 \times 1.5 + 900 \times 3 + 4600 \times 4 = 23350$ ft-lb.

Factor of safety: $F_s = M_R/M_o = 23350/8504 = 2.75$.

Since it is >1.5. It is acceptable.

2. Check Soil Bearing Capacity

Total weight of retaining wall: $W = 1500 + 900 + 4600 = 7000$ lbs.

The distance from center of resultant to toe,

$$X = (M_R - M_o)/W = (23350 - 8504)/7000 = 2.12 \text{ ft.}$$

Eccentricity: $e = 6/2 - 2.12 = 0.88$ ft $< 1/6$ width of footing, 1 ft.

Maximum and minimum footing pressure:

$$Q_{max} = \frac{W}{B}\left(1 + \frac{6.e}{B}\right) = \frac{7000}{6}\left(1 + \frac{6 \times 0.88}{6}\right) = 2192 \text{ psf}$$

Since it is < 3000 psf. It is acceptable,

$$Q_{min} = \frac{W}{B}\left(1 - \frac{6.e}{B}\right) = \frac{7000}{6}\left(1 - \frac{6 \times 0.88}{6}\right) = 141 \text{ psf}$$

3. Check Sliding Stability

Driving force: $P_h = g\, K_a H^2/2 = 2319$ lbs.

Friction resistance at bottom of footing: $F_r = m\, W = 0.5 \times 7000 = 3500$ lbs.

Rankine's passive earth coefficient: $K_p = \tan(45 + f/2)2 = 3$.

Height from top of backfill soil to toe: $h = 2$ ft.

Passive resistance at toe: $P_p = g\, K_p H^2/2 = 115 \times 3 \times 22 = 690$ lbs.

Factor of safety against sliding:

$$F_S = (F_r + P_p)/P_h = (3500 + 690)/2319 = 1.81.$$

Since it is >1.5.

It is acceptable.

Permissions

Index

www.ingramcontent.com/pod-product-compliance
Lightning Source LLC
Chambersburg PA
CBHW062001190326

41458CB00009B/2936